ENCOUNTER IN
RENDLESHAM
FOREST

ALSO BY NICK POPE

ENCOUNTER IN RENDLESHAM FOREST

The Inside Story of the World's
Best-Documented UFO Incident

NICK POPE with
JOHN BURROUGHS, USAF (Ret.),
and JIM PENNISTON, USAF (Ret.)

THOMAS DUNNE BOOKS
St. Martin's Griffin ≋ New York

THOMAS DUNNE BOOKS.
An imprint of St. Martin's Press.

www.thomasdunnebooks.com
www.stmartins.com

Designed by Steven Seighman

The Library of Congress has cataloged the hardcover edition as follows:

Pope, Nick, 1965–
    Encounter in Rendlesham Forest : the inside story of the world's best-documented UFO incident / Nick Pope with John Burroughs, USAF (Ret.), and Jim Penniston, USAF (Ret.) — First edition.
        p. cm.
    Includes bibliographical references and index.
    ISBN 978-1-250-03810-4 (hardcover)
    ISBN 978-1-250-03811-1 (e-book)
    1. Unidentified flying objects—Sightings and encounters—England—Rendlesham Forest. I. Burroughs, John, (former Air Force sergeant), author. II. Penniston, Jim, author. III. Title.
    TL789.6.G7P67 2014
    001.94209426'4—dc23

                                                            2013046840

ISBN 978-1-250-06331-1 (trade paperback)

St. Martin's Griffin books may be purchased for educational, business, or promotional use. For information on bulk purchases, please contact the Macmillan Corporate and Premium Sales Department at 1-800-221-7945, extension 5442, or write to specialmarkets@macmillan.com.

First St. Martin's Griffin Edition: March 2015

D 10

To Elizabeth

With all my love, forever

CONTENTS

ACKNOWLEDGMENTS

This book owes its existence to a large number of people.

First, I would like to thank my literary agent, Andrew Lownie. I have been agented (I'm not sure if that "word" will get past my editor!) by Andrew for nearly twenty years now and I know that as an author himself, he "gets it." He has consistently given me excellent advice and worked tirelessly to get the best possible deals for my books. In a world where the word is sometimes misused or abused, he has genuinely steered me to "bestseller" status.

I would also like to thank all those at St. Martin's Press who have worked so hard on this project, helping to ensure that this final product is as polished as possible. I would particularly like to thank my editor, Peter Wolverton, and his assistant, Anne Brewer, for their vision, words of wisdom, and effort. They ensured that the editorial process was an absolute pleasure and added value at every step. Thanks to the graphics department for the striking and evocative front cover and to all those in sales, marketing, and PR who have put so much effort into making this book a success.

I should also like to pay tribute to other authors, journalists, broadcasters, and researchers who have helped us uncover the truth about the Rendlesham Forest incident. Their hard work and perseverance helped blaze a trail and bring us closer to the truth. Most of those concerned have been acknowledged in the text itself, but I should like to highlight in particular John Alexander, Keith Beabey, Georgina Bruni, Jasper Copping,

Chuck de Caro, Darren DeBoy, Linda Moulton Howe, James Fox, Timothy Good, Neil Henderson Leslie Kean, and Salley Rayl.

Thank you to astronaut and all-American hero Edgar Mitchell and to former Canadian defense minister and deputy prime minister Paul Hellyer for sharing your views about this incident. Posthumous thanks to the former UK Chief of the Defence Staff Lord Peter Hill-Norton for his indefatigable quest to uncover the truth about these events.

I am also grateful to all those who helped with the "Rendlesham Code," bringing their respective energies, insights, and expertise to bear on this most intriguing aspect of the case. Particular thanks are due to Nick Ciske, Joe Luciano, Gary Osborn, and Kim Sheerin.

This book would not have been possible without the cooperation, advice, and assistance of a large number of men and women based at Bentwaters and Woodbridge at the time of the Rendlesham Forest incident. I am grateful to all concerned, but in particular to John Coffey, Charles Halt, Charles (Chuck) Heubusch, Monroe Nevels, and Gordon Williams. I offer my particular thanks to all those who served in the 81st Security Police Squadron who, along with other colleagues, were caught up in these events. I hope you get the answers you are looking for.

Thank you to various staffs at the UK Ministry of Defence, the US Department of Defense, and the UK National Archives for various technical assistance and thank you to various other people within government, the military, and the intelligence agencies who have helped with information, insights, or support but who cannot be named.

My heartfelt thanks go to Pat Frascogna for being part of the team and for his excellent overview (reproduced in full in chapter 17) of the ongoing work to use the Freedom of Information Act to push various agencies to reveal what they know about the Rendlesham Forest incident. Pat has been a tireless campaigner on behalf of John Burroughs and Jim Penniston and has given most generously of his time and expertise.

Thank you to Patricia Bullock Williams for her painstaking analysis of the audiotape made on the night of the second encounter in Rendlesham Forest and her meticulous transcript.

My most profound thanks, of course, are reserved for my two coauthors. John Burroughs and Jim Penniston were the two USAF personnel at the heart of this incident. In the years since the encounter they have suffered Post-Traumatic Stress Disorder and other health problems and have been frustrated as a mixture of half-truths, rumors, exaggerations, and lies have been told about this case. These two men served their country with honor and distinction. It is high time that their loyalty was repaid.

I would also like to offer a heartfelt thank-you and an apology to anybody whom I have inadvertently forgotten to mention by name here. I am no less grateful, despite the omission.

Finally and most important, I am grateful to my wife, Elizabeth Weiss, for all her support. I love you.

GLOSSARY OF ABBREVIATIONS

ADGE—Air Defence Ground Environment

AFOSI—Air Force Office of Special Investigations

CAA—Civil Aviation Authority

CDS—Chief of the Defence Staff

CIA—Central Intelligence Agency

CINCUSAFE—Commander in Chief US Air Forces in Europe

DARPA—Defense Advanced Research Projects Agency

DAS—Directorate Air Staff

DGSTI—Director General Scientific and Technical Intelligence

DIA—Defence Intelligence Agency

DIS—Defence Intelligence Staff

DoD—Department of Defense

DS8—Defence Secretariat 8

FAA—Federal Aviation Administration

FOIA—Freedom of Information Act

IFF—Identification Friend or Foe

LE—Law Enforcement

MoD—Ministry of Defence

NASA—National Aeronautics and Space Administration

NBC—Nuclear, Biological, Chemical

NCND—Neither Confirm Nor Deny

NCO—Non-Commissioned Officer

NSA—National Security Agency

PRP—Personnel Reliability Program

PTSD—Post-Traumatic Stress Disorder

PQ—Parliamentary Question

RAF—Royal Air Force

SAS—Special Air Service

Sec(AS)—Secretariat (Air Staff)

SME—Subject Matter Expert

SOP—Standard Operating Procedure

SP—Security Police

SRAFLO—Senior Royal Air Force Liaison Officer

UAP—Unidentified Aerial Phenomenon

USAF—US Air Force

USAFE—US Air Forces Europe

VA—US Department of Veterans Affairs

WSA—Weapons Storage Area

INTRODUCTION

Almost everyone has heard of the Roswell incident. Even if someone has no interest in UFOs and no particular beliefs on the subject, mere mention of "Roswell" will likely lead to some recognition: "Oh yes," people will say, "the place where a UFO is supposed to have crashed in 1947." Some believe it, others don't, and a whole bunch of people are undecided, but the point is, just about everybody has heard the story. It goes further than this. Roswell not only is the subject of countless books and TV documentaries but also has become part of pop culture. When a video emerged in 1995 purporting to show the dissection of an extraterrestrial from the crash, the so-called alien autopsy film was front-page news all around the world. Roswell was name-checked in the sci-fi movie *Independence Day*, and the town of Roswell, in New Mexico, boasts a UFO museum and puts on an annual UFO parade where thousands turn out.

The problem is, whatever happened at Roswell took place well over sixty-five years ago. The last surviving witness, Jesse Marcel Jr., died on August 23, 2013. He was ten years old back in 1947 when, as he recounted the story, his father—an Air Force intelligence officer—woke him up in the middle of the night and showed him strange debris, telling him it was from a spaceship and that he'd never see anything like it ever again.

A US infantryman once wrote: "No war is really over until the last veteran is dead." Once this has happened, a war passes from living memory into history. So, with the passing of Marcel, Roswell has become part of history. This being the case, how can there be a meaningful discussion about UFOs if the best-known and most often-cited UFO incident took place so long ago and no contemporary witnesses remain to shed any light on what happened? Surely there's a more recent UFO incident that can become the new focus of the debate? Can't we do better than dragging up Roswell yet again? The good news is, we most certainly can! This book tells the story of just such a UFO incident, and unlike Roswell, the witnesses are very much alive—and ready to tell their stories.

Simply put, the Rendlesham Forest incident is by far the best-documented and most compelling UFO incident ever to have taken place. It took place over a series of three nights, in a forest that lay between two U.S. Air Force (USAF) bases in England. This was not some vague "lights in the sky" sighting—the UFO actually landed. But unlike Roswell, where it's claimed that the craft crashed, with the wreckage (and maybe bodies, too) being spirited away by the military, this UFO was seen to take off again.

Unlike Roswell, where there are lots of stories and rumors but few hard facts, this is a case where there is an abundance of evidence. This book tells the complete story of this incident from the landing on the first night to the subsequent return of the UFO as witnessed by the skeptical Deputy Base Commander and a group of tough, skeptical USAF military personnel.

As compelling as testimony is at this level—multiple witnesses; military personnel, some of long seniority and high rank—the book you are about to read will do far more than set out eyewitness accounts. We will present radar data and detail physical evidence to back up the statements, including information about abnormally high radiation levels found at the landing site. All such claims are backed up by verifiable sources, such as formerly classified UK Ministry of Defence (MoD) documents obtained by informed and carefully targeted use of the Freedom of Information Act.

At the heart of this book lies the testimony of the two witnesses most closely involved in this incident, John Burroughs and Jim Penniston.

John Burroughs served for twenty-seven years in active and reserve duty in the USAF. At the time of the incident he was an airman first class, assigned as a Security Police/Law Enforcement patrolman at RAF Bentwaters. His numerous deployments included various assignments to the

Middle East and some of the details remain classified to this day. Burroughs left the Air Force in 2006, as a Law Enforcement supervisor.

Jim Penniston is the more senior of the two men, having joined the USAF in 1973. In 1980, when the incident occurred, he was a staff sergeant: a senior and experienced Non-Commissioned Officer (NCO) who held the post of Senior Security Officer. He was in day-to-day charge of base security and was responsible for the protection of the war-fighting resources at the base. Penniston served throughout Operations Desert Shield and Desert Storm, though as with Burroughs, many details of his military career are still classified. He retired from the USAF in 1993.

While there are dozens—if not hundreds—of USAF personnel who were involved in some aspect of the Rendlesham Forest incident, it is Burroughs and Penniston who are central to this story. They are the two individuals who, responding to reports of a possible security incident on December 26, 1980, found themselves face-to-face with the landed UFO. They are the ones for whom the ufology/sci-fi phrase "close encounter" was to become a chilling reality.

John Burroughs, Jim Penniston, and I have collaborated on this book, and while I have taken on the role of lead writer (somebody had to draw the short straw and sit down in front of the computer!), this is a joint venture in every sense. We have worked together on this project for many months, pooling knowledge, swapping insights from our respective areas of expertise, brainstorming, debating theories, and generally trying to make sense of the wealth of information that we have amassed on the Rendlesham Forest incident.

Even when not featured directly in some parts of the narrative, the inside knowledge of Burroughs and Penniston is what has enabled this story to be told. Aside from relaying what they experienced firsthand, Burroughs and Penniston have located other witnesses and persuaded them to place their accounts into the public domain. More generally, Burroughs's and Penniston's in-depth knowledge of the bases, the personnel, the mission, and the standard operating procedures (SOPs) has enabled us to separate fact from fiction and to put the events into their correct politico-military context.

This book tells the story not just of these spectacular UFO encounters but also of what happened afterwards. In many ways, this is just as compelling as and even more disturbing than the UFO sightings themselves. The aftermath of the incident saw not just an attempt to remove evidence and silence witnesses but also official debriefings where threats were made and where narcotics and hypnosis were used first to uncover, then to distort, and finally to bury the truth. John Burroughs and Jim Penniston, in particular, found themselves in the eye of this storm.

One particularly dark and disturbing aspect of this story relates to the health problems that Burroughs and Penniston are now suffering—problems that they attribute to their close proximity to the landed craft. Of particular concern, given a Defence Intelligence Staff (DIS) assessment of readings taken at the landing site by the Disaster Preparedness Officer, using a Geiger counter, is the possibility that they were irradiated by the object. This would certainly explain some of the problems the two men have experienced and this is why they are seeking to obtain their military medical records. This, in conjunction with clarification from the US military on what they encountered, would enable Burroughs and Penniston to obtain proper diagnosis and treatment.

Unfortunately, what should be a simple administrative procedure—obtaining two medical files—is proving to be a bureaucratic nightmare. Burroughs's and Penniston's medical records appear to be held by a little-known classified-records section in the Department of Veterans Affairs. Freedom of Information Act requests, a threatened lawsuit, and engagement with various congressional representatives have all been brought to bear on what is an ongoing situation.

This is not a UFO book in any conventional sense. Most UFO books—whether or not the authors come out and say so—are written to make a case that UFOs are extraterrestrial spacecraft. A much smaller number of UFO books are written with the intention of disproving the so-called extraterrestrial hypothesis. Both types of books are conclusion led. The authors peddle their own dogma, be it that of true believer or die-hard debunker. Data are used selectively, erroneously, or downright dishonestly in such a way as to convince readers either that we're being visited by space aliens or, conversely, that such a thing is impossible.

This book takes a fundamentally different approach. It's data led and we place into the public domain a vast amount of information, most of it published here for the first time. This comprises testimony from the witnesses most closely involved with these events, information from newly declassified government UFO files, and insights from our respective positions within the UK MoD and the USAF. While we give readers an insider's perspective of the secretive government and military world in which this story unfolds, we leave it to readers to assess the information and decide for themselves what they think.

What follows is the inside story of a UFO incident that's bigger than Roswell, but it's also a very human story of two men on a quest. What happened to John Burroughs, Jim Penniston, and others in December 1980 is bizarre and terrifying enough in its own right. But what happened next is even more disturbing, as they struggled to come to terms with a life-changing, paradigm-shifting event, as they searched for answers, and as they took on a government that on the one hand says UFOs are a non-issue but on the other behaves as if the topic was as highly classified as any subject in government.

This is a timely book. Interest in UFOs is at an all-time high. Programs on the subject are a constant fixture on TV schedules. Perhaps network executives sense a quickening of pace here, with a seemingly constant stream of stories about new astronomical discoveries that take us closer and closer to finding planets like Earth. Perhaps they realize that few questions are bigger, more profound, more controversial, and more impactful than "are we alone in the universe?" and "are we being visited?"

This book is timely for another reason. After a period when such things either were not thought about much or seemed to be accepted, concerns over government secrecy have risen to near the top of the political and media agenda. Fueled by sites such as Julian Assange's WikiLeaks and by the revelations of whistle-blowers like Chelsea (formerly Bradley) Manning and Edward Snowden, the secret state is big news. Whether it's the National Security Agency's (NSA's), PRISM program, drones, or worries over security of personal data, US citizens realize the time is right to have an informed debate on the subject. The story of the Rendlesham Forest incident raises these very issues. On the one hand, we have the natural desire of John Burroughs and Jim Penniston to find out what happened to them. Open government and freedom of information may help them. But on the other hand, it seems clear that whatever happened to Burroughs, Penniston, and their other military colleagues, someone in government wants it to stay secret and is prepared to go to great lengths to ensure things stay that way. Is this heavy-handedness, or are there legitimate reasons why the truth about these encounters must never be made public?

With these brief remarks made, it's time to begin our journey of discovery. It's a journey that begins in the darkness of an English forest and ends in the White House.

ENCOUNTER IN RENDLESHAM FOREST

1. GROUND ZERO

In the early hours of December 26, 1980, there's little outward sign of activity at the twin US Air Force bases of Bentwaters and Woodbridge. The bases lie a few hundred yards from each other. Between them lies Rendlesham Forest. The twin bases are much as you'd expect military bases to be—though they hide a secret that would horrify most people in the area, who were blissfully unaware of the bases' true mission. Bentwaters and Woodbridge lie in the sleepy county of Suffolk, on the cold, exposed coast of the East of England. For most of the young men and women here, it's their first experience of a foreign country. To soften the blow, the bases have all the comforts of home—a bar, a burger joint, and other stores—they're like small American towns nestling in the heart of the English countryside.

The events started when Airman First Class John Burroughs spotted strange lights in the forest. Burroughs had been patrolling Woodbridge and was close to the East Gate (sometimes colloquially referred to as the back gate). The lights were due east from that location and were red and blue. The red light was above the blue light and they were flashing on and off. Burroughs altered his supervisor, Staff Sergeant Bud Steffens. Both men watched in amazement as despite being familiar with the area (Burroughs had been based there for seventeen months), they had never seen anything like it before.

Their first thought was that an aircraft might have crashed in the forest. Not one of the A-10s (there was no military flying activity on the night concerned) but perhaps a civilian light aircraft. Their first and most basic instinct, of course, was to investigate and to render assistance if it was needed. Also, there was the question of security. If some unexplained activity was going on so close to the twin bases, was there a threat—actual or potential?

Burroughs unlocked the combination lock on the East Gate and he and Steffens drove out of the gate and a couple of hundred yards or so down to a small public road. There they turned right and drove another ten or twenty yards, before reaching a left-hand turning where a small track led into the forest. At this point, a white light was visible, in addition to the red and blue lights. This white light was particularly odd and at one point appeared to be coming closer to them, down the small track. The color, configuration, and movement of the lights were like no aircraft or vehicle they were familiar with.

Despite the urge to press on, they realized they needed to call in this incident, so drove back to the East Gate, where they phoned in a brief report—using the landline in the guard shack, as opposed to their pocket radios, which were known to be insecure and susceptible to scanners, which could be used to pick up conversations.

Burroughs spoke to the Law Enforcement (LE) duty desk sergeant, Sergeant "Crash" McCabe, and explained what was happening. McCabe wasn't sure what to make of this and briefly wondered whether some sort of practical joke was being played on him. He asked to speak to Steffens, who confirmed what Burroughs had said was true and that he, too, had witnessed the strange lights. McCabe also suspected that an aircraft crash might have occurred and called through to Central Security Control, passing the problem to Staff Sergeant John Coffey.

Coffey called the on-duty flight chief at RAF Woodbridge, Staff Sergeant James (Jim) W. Penniston. Penniston wasn't briefed on the nature of the situation but was told to proceed with his rider, Airman First Class Edward N. Cabansag, driving to the East Gate with blue lights on to rendezvous with "Police Four" and "Police Five"—the call signs for Burroughs and Steffens. This was highly unusual, and Penniston was somewhat flustered and annoyed that he wasn't briefed on what to expect but was simply told to rendezvous with Burroughs and Penniston at the East Gate, where he'd be told the nature of the situation. This departure from standard procedure was one of the first indications that this was a highly unusual and sensitive situation. It also raises the possibility that at least some people in the chain of command already knew more about this than they were saying or had been instructed not to give details of the situation over the communications systems. Otherwise, why not simply say to Penniston something along the lines that a patrol was investigating a possible aircraft crash in the forest? Coffey could have been even vaguer and used a phrase such as "possible security situation" or that phrase so beloved by police all around the world, "an incident." There was no problem in terms of jurisdiction and USAF personnel were certainly allowed to patrol off-base (outside the wire, as it's called in the military) in a wide range of circumstances.

While there's some confusion over the exact time, Jim Penniston recalls that it was just after midnight.

Burroughs and Steffens were still waiting at the East Gate when they were joined by Penniston and his driver, Airman First Class Edward Cabansag. They quickly briefed Penniston, and once again the view of the experienced flight chief was that this must have been an aircraft crash. But it was the middle of the night, at Christmas, and there was certainly no military aircraft activity. And while the possibility of a civil aircraft was still being considered, nobody had heard an explosion or any sounds. And that's when Steffens made an odd remark that caught everyone's attention:

"It didn't crash. It landed."

Despite that disconcerting observation, Penniston felt the aircraft crash theory was still the most likely explanation and with this in mind radioed Central Security Control and asked to speak to the overall flight chief for both bases, Master Sergeant J. D. Chandler.

If all this "A calls B, who then checks with C" procedure seems somewhat labored, especially in a situation where those involved might have been dealing with an aircraft crash, there are three important points to bear in mind. First, while looking at a written account might lead readers to think valuable time was being wasted, most of the actions set out previously are relatively quick and easy ones and—in the case of the telephone calls and radio conversations—take only seconds. Second, the military is a notoriously hierarchical organization where everybody is very rank conscious; in such a culture, clearing a non-routine action with your supervisor, or at the very least informing him or her, is rather more important than it is in most civilian organizations—getting your top cover, as it's sometimes referred to. Finally, the security police and law enforcement specialisms tend to be very process driven in many respects. Initiative is still encouraged, but a lot of tasks are performed by following a set procedure that's learned by heart and then constantly tested and reinforced through training.

With the aircraft crash theory in mind, Chandler checked the position with regard to aircraft activity with the control tower at Bentwaters. Somebody in the control tower checked the radar and also placed calls to RAF Bawdsey, RAF Watton, and Heathrow Airport in London. The key piece of information that came back was that a "bogey"—or "bogie"—(defined by the USAF as "a radar or visual air contact whose identity is unknown") had been tracked around fifteen minutes previously but that the target had been lost when it disappeared from the radar screen directly over the Woodbridge base. Chandler contacted the shift commander and gave Penniston the OK to continue the investigation. Perhaps because he sensed trouble or knew something was amiss, Penniston requested backup. In response to this request, Chandler decided to come out himself.

With the somewhat troubling piece of news about the radar return having been relayed to them, Penniston, Burroughs, and Cabansag drove out into the forest to resume the investigation that Burroughs and Steffens had started shortly beforehand. There is some confusion about why Steffens didn't go out into the forest. One possibility was that with personnel about to go off-base, weapons needed to be left with someone—though, in fact, weapons can be taken off-base in some circumstances, e.g., where an immediate and serious security threat is perceived. Indeed, there's a suggestion that some of those who went into the forest didn't leave their weapons behind, even if they should have!

Penniston, Burroughs, and Cabansag took the same route as had been taken before: they drove the couple of hundred yards or so from the East Gate to the small road that ran through the forest. They turned right, drove for a few yards, then turned left down a small track that led deeper into the forest. These tracks are not proper roads and are very narrow and bumpy. You can't drive a vehicle—even a sturdy one like a military Jeep—that far down them, so after maybe no more than fifty yards or so the men had to stop the car and proceed on foot.

As they advanced into the forest with—as Cabansag described it—"extreme caution," all three men could see the strange lights. Cabansag described them as being "blue, red, white, and yellow."

Though not formally classified as such, the event clearly was now being treated as a potential security situation. While the only theory that had been discussed so far was that this was a potential crash of a light aircraft, the facts simply didn't add up. And by this time, none of those involved thought this was what they were dealing with: the obvious proof of this is that nobody had called for an ambulance (or even a first-aid kit!) or called out the fire brigade. Another clue was Cabansag's admission that they proceeded with "extreme caution"—hardly the actions of a patrol engaged on an urgent search-and-rescue mission.

By this time, the backup they'd requested had arrived. This consisted of Master Sergeant Chandler (whom Penniston had spoken to earlier) in another vehicle. There's confusion about who arrived first. Chandler says that when he arrived Penniston, Burroughs, and Cabansag "had entered the wooded area just beyond the clearing at the access road," while Cabansag said that Chandler was already "on the scene." Such inconsistencies may seem minor, but they're indicative of something wider, because while four men were in the forest that night, all came back with different memories of what happened next.

At about this time, all four men's radios began to malfunction. Or rather—given the small chance that four separate radios would simultaneously malfunction—something began to interfere with communications, which seemed only to be working over a short distance. To deal with this, the four men adopted a low-tech solution and set up a radio relay. In other words, Chandler stayed with the parked vehicles and from there was able to relay messages between the men who went deeper into the forest and his colleagues back on the base, in Central Security Control. Cabansag went forward, but when he and Chandler could hardly hear each other he, too, stopped, leaving only John Burroughs and Jim Penniston to push forward to try to find the source of the lights.

Burroughs and Penniston soon realized there was something seriously wrong. The air was filled with static electricity and the hairs on their arms and on the backs of their necks were standing on end. It was difficult to walk properly and they described the experience as being akin to wading through deep water. All the time, the lights were ahead of them, getting brighter and more clearly defined as they ventured deeper into the forest, closer to whatever was out there.

Up ahead was a small clearing, brightly illuminated. They had reached their goal. Suddenly, as they approached, there was a silent explosion of light. They instinctively hit the ground, fearing they'd be hit by debris from the bright flash of light. Penniston, seeing no apparent harm from the immense flash of light, stood up, and what came clearly into view was clearly nothing to do with an aircraft crash.

Penniston looked to his right and saw Burroughs engulfed in a huge beam of light, which appeared to be coming from above. The light encompassed Burroughs. Then Penniston saw that what had first appeared to be a sphere of light in front of him had dissipated and now had the appearance of a craft of some sort.

Staggered, Penniston took stock of the situation. In the clearing was a small, metallic craft. It was about three meters high and maybe three meters across at the base. The craft was roughly triangular in shape and appeared to be either hovering just above the ground or perhaps resting on legs at each edge of the object, as if it was on a tripod, like a lunar landing module (only with three legs and not four). It had a bank of blue lights on its side and a bright white light on the top. There was no sound whatsoever.

As Penniston approached the object he saw strange symbols on the side. They were unlike anything he'd seen before, and the nearest match he could come up with was ancient Egyptian hieroglyphs. Penniston had the presence of mind to take a number of photographs and sketch both the craft and the symbols in his police notebook.

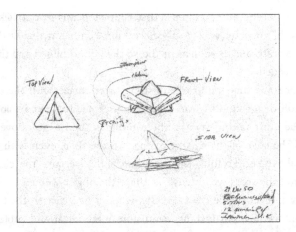

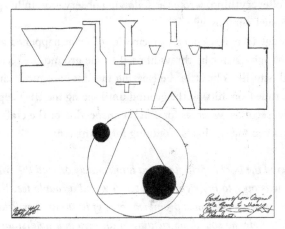

Finally, Penniston plucked up the courage to touch the object. It felt hard and smooth. This, combined with the look of the hull, close-up, made him think of a smooth, opaque black glass. He then moved to touch the symbols. He recalls the sensation thus: "The skin of the craft was smooth to touch. Almost like running your hand over glass. Void of seams or imperfections, until I ran my fingers over the symbols. The symbols were nothing like the rest of the craft, they were rough, like running my fingers over sandpaper."

As Penniston touched the symbols, the white light on top of the craft flared up and became so intense that Penniston was fear struck and temporarily blinded by what was before him. Penniston removed his hand from the craft, and as soon as he did so the light dimmed and the sense of panic receded.

After some time, and to Penniston's utter amazement, the craft lifted slowly off the ground. Again, everything seemed to move in slow motion, with the craft taking two or three minutes to rise up above the trees around the edge of the small clearing. All the time, even with the object rising above the ground, there was no noise. Because the clearing was small and the trees were dense, at times the object seemed as if it had to maneuver through the trees. Finally, when it had cleared the trees, it accelerated away in an instant. Penniston, methodical and professional in the face of everything, wrote the following observation in his police notebook: "Speed—impossible."

Burroughs has few coherent memories of what happened after the explosion of light. After he threw himself to the ground, he recalls seeing a red, oval, sun-like object in the clearing but does not recall the craft. For him, the time from hitting the ground until seeing the UFO depart seemed like a few seconds, whereas for Jim the inspection of the craft took many minutes. Even today, this is troubling for Penniston:

I entered the bubble field (the area immediately around the craft) first; John was over to my right about ten feet and a couple feet back. The silence was then the most prominent part of it; the area or field seemed dead; the air: no sound; no rustling of air or wind; no distant sounds, no animals or nothing—a dead silence. A strong static on clothes, hair and skin—being pulled toward the light. Then dissipated—I was alone. And from John's perspective, he has no memory. John is standing still and motionless. I yelled at him, of course. No reaction; he does not move. He, of course, cannot hear me and I then turn and focus on the craft and the matter of security for the bases. It has been always the case that John does not have a memory of this. But when we were being debriefed and writing statements in Colonel Halt's office, less than 72 hours after the first night, John in his statement (which was hand written) has the drawing of the craft he saw with me. This has always made me wonder about John's memory. Why could he do this within 72 hours and today has no memory? Definitely food for thought!

Food for thought indeed. Especially when combined with one cryptic comment Burroughs made when pressed on where he was while Jim examined the craft and why their memories are so different at this point: "The only possibility is that I was in the light when he was doing his examination."

Penniston and Burroughs—still in a state of considerable shock—attempted to relocate the UFO and had a number of further sightings of strange lights on the horizon. At one point the object was so close they thought it was going to land again. But it didn't and the UFO eventually departed to the east, out over the coast.

Still confused and disorientated, they eventually decided to make their way back out of the forest. As they did so, they passed back through the small clearing where they'd had their encounter. Still trying to process what had happened to them, they looked around. Perhaps if they found nothing, they'd somehow convince themselves that it had been some sort of shared hallucination.

It was not to be. In the very center of the clearing, on the hard, frozen ground, were three indentations. They were recent. Something heavy—probably weighing several tons, judging by the hardness of the ground and the depth of the indentations—had clearly been resting there. When they looked more closely, they saw that if they drew an imaginary line between the three indentations the shape formed would be a near-perfect equilateral triangle.

As further confirmation, they noticed that branches had been snapped off the trees around the edge of the clearing, where the object had smashed its way in from above and then done the same on its way out. It would sound absurd, were it not for the fact that this was precisely what Penniston had just witnessed.

They left the clearing and rendezvoused with Cabansag, Chandler, and six other security force members, before making their way back to RAF Bentwaters, the main operating base. When they arrived back, they found that they'd been gone much longer than they realized. This, coupled with the fact that they'd been out of radio contact, had caused near panic in certain quarters. Indeed, a search party had been on the point of going out to look for them. Penniston tried to convince himself that adrenaline would explain the time discrepancy, but their watches told a different story. He explains: "I suppose anything is possible with this time discrepancy. I believe it is more than likely that within the affected area around the craft there was a distortion of some kind, which based on the missing time from our watches indicates this, by them running forty-five minutes slow. We were definitely affected by this phenomenon in a physical way, including the machinery we wore (watches)."

Burroughs confirms both the "missing time" and the exact figure: "The fact is our watches were behind and the shift commander said we were missing for 45 min."

There were some formalities to go through. Weapons had to be returned and signed for and a hurried debriefing—the first of many—was carried out. Though they didn't realize it at the time, dozens of other military personnel at the twin bases had seen the strange lights and had been watching from a number of vantage points, including the control tower at Bentwaters. Everyone wanted to know what had happened to Penniston and Burroughs, but for them the immediate aftermath of their encounter was a blur. They just wanted to get off-duty, go back to their beds, sleep, and forget. They soon got their wish, but if they thought their ordeal was over, they were sadly mistaken. It had only just begun.

Burroughs and Penniston were, it should be said, skeptical about UFOs. In a sense, they still are, despite everything. Penniston set out his views this way:

My thoughts are simply that 99 percent of all so-called UFO sightings can be explained by people with a knowledgeable background or aerial training to reporting such things for exactly what they are. Their UFO is an observation of the following type of possibilities: as a manmade object, star/planetary body, or other natural occurring phenomena, all completely identifiable by a trained observer. There are also people who seem to have some physiological issues which are in my opinion manifested exponentially when they see things they can't explain. I feel it is a natural state for those with tainted objectivity within their thinking and can have easily have occurred with other sightings. With that entirely aside with the 99 percent, this leaves the remaining one percent. It is this percent I believe is the truly unknown—a conclusion I made after I left the forest that night. This is the very reason I am troubled by the events of December 26, 1980. I went into the forest with what I just said being the case. Then I left the forest with the "One Percent Factor" raining all over me. I had no answers for what clearly created conditions, effects and the presence of an unknown craft with technology that cannot be replicated even today. So how does this all play? Simply, one percent of UFO sightings are unknown.

Burroughs is more succinct, though his dismissive remarks about UFOs betray, perhaps, a sense of unease alongside his skepticism: "I never spent any time thinking about any of that. The only feeling I ever had was I hoped I never had to walk a mile in their shoes."

Penniston went on to sum up the transformative nature of the experience like this: "What I once believed is no more and what I've witnessed defies all that I have ever imagined. I am truly in awe over the whole incident and no-one can fully understand the magnitude of such an event unless you were there."

For jurisdictional and legal reasons, it was important that the American military notified the British authorities that they were going off-base. The usual way in which this was done was by notifying the local police. This action had been taken at about 4:00 am by Airman First Class Chris Armold on the LE desk. In a message to Suffolk Police he wrote: "We have a sighting of some unusual lights in the sky, have sent some unarmed troops to investigate, we are terming it as a U.F.O. at present."

Two police officers responded and briefly searched the area. They found nothing, though their inquiries revealed the fact that strange lights had been witnessed over large parts of southern England earlier in the night.

A later entry in the Suffolk Police log provides the first known documentary evidence showing that a landing *had* taken place. This intriguing log entry reads as follows: "We have had a call from the L.E. at Bentwaters in reference to the U.F.O. reported last night. We have found a place where a craft of some sort seems to have landed."

One of the other central figures in this story is the Deputy Base

Commander, Lieutenant Colonel Charles I. Halt. Halt was a thorough and careful man. He had wide-ranging responsibilities for security, policing, law enforcement, and a large number of administrative functions at the twin bases. Halt also liked to "walk the ground." That's to say, he often took an in-depth look at some of the areas for which he had responsibility. Sometimes, for example, he'd sit with the fire department personnel, talk to the cooks, or ride with the cops on patrol. Some officers wouldn't have gotten that close to the enlisted men and women under their command (many officers are more detached and keep a "professional distance" from those under their command) and some of these more junior personnel were pretty nervous about being so closely scrutinized by one of the most senior officers at the twin bases. Halt, however, felt that if he was to do his job properly, he needed to get down into the weeds and understand every aspect of what was going on. He regarded it as the best way to get what the military call ground truth—not what people tell you is going on but what's *actually* going on. Burroughs acknowledges this trait but was not keen on it, clearly feeling a little "over-supervised": "He was high speed; would ride around with LE to get a feel of what was going on at the base. He rode with me a couple of times, always was getting in our way."

Penniston clearly had great respect for Halt:

Colonel Halt is what I call "An Enlisted man's Colonel"; he valued the Non-Commissioned Officers (NCOs) that he commanded. He valued their knowledge, skills, assessments, opinions, judgments and the people themselves, as a valuable and key part of the United States Air Force mission. Colonel Halt is an officer who truly believes you are only as good as people you command. From Major Command evaluations to local evaluations, the Colonel believes it was the NCO corps that made it all happen. Then his conduct in regards to the Rendlesham Forest Incident, well, he was only following orders, and he stretched those orders as far as he could without jeopardizing his career.

The morning after the UFO encounter, at about 5:00 am, Halt came on duty and headed for the LE desk. There was some chatter and laughter, which abruptly stopped as he entered the room.

"What's going on?" Halt asked.

"Penniston and Burroughs were out last night chasing UFOs, sir," Sergeant McCabe replied.

McCabe was writing up the LE blotters—the logs on which anything significant that occurred on the shift were recorded. The purpose of these logs was twofold—they were a useful part of the handover process to the next duty shift and they also provided a source of raw data that was invaluable if a question was raised later, when memories had faded.

"Put it in the blotter," Halt ordered.

It later transpired that accounts of what had happened were written up not just in the Law Enforcement blotters but also in the Security blotters and that an Air Force Form 1569 (Incident/Complaint Report) was completed by the security controllers at Central Security Control.

Everyone was sensitive about using the loaded term "UFO," so Halt suggested using a vague phrase such as "unexplained lights" and maybe making reference to the theory concerning a possible light aircraft crash. Halt was to use the same phrase, "unexplained lights," later, when reporting the incident formally to the British government, via the Ministry of Defence (MoD). So already, just hours after the incident, the matter was being played down. But perhaps more sinister forces were at work. Later that day, Halt became aware that the encounter Penniston and Burroughs had had was something far more tangible than a mere "lights in the sky" UFO sighting. Halt moved quickly to review not just the LE blotter (which he'd instructed be written up in vague terms) but also the Security blotter, which was likely to have the best and most accurate contemporaneous account of what actually took place.

Halt's plan was frustrated. Somehow, somebody had removed both blotters *and* the incident report, with nobody on duty being able to explain how. Staff Sergeant Coffey recalls "my Blotter was pulled and classified SECRET by the Base Commander [Colonel Ted Conrad]." Penniston offers this view of the situation:

The removal and classifying of the security blotters (AF Forms 53) and AF Forms 1569s (Incident and Complaint Report) were part of the containment process initiated by others, outside the Base Command. Colonel Halt unknowingly asked the Desk Sergeant to include the first night's information that they had already omitted on the

morning after. I think it disturbed him when he became aware the se-
curity blotters and 1569s had already been classified and pulled. With
the removal of the blotters and 1569s, [without Halt's knowledge] it
was much easier to put out a cover/containment story about the nights
in question.

Burroughs takes a more conspiratorial view here and offers a view on
where the material was sent: "That, along with all the other missing documents, shows that this incident was classified early on. Also, there is no
way they just disappeared. My guess is they were sent to Germany [HQ
United States Air Forces Europe (USAFE)] and that the State Department
got a copy too."

The blotters were never found, so right from the very outset we have not
just a spectacular UFO encounter but also strong evidence of a conspiracy.
Standard procedures were being ignored and evidence was somehow being
removed from right under the noses of people for whom security was a way
of life. It's difficult to see how the blotters and the incident report could
have been spirited away without one of the key players being involved in
some way. The alternative—that outsiders were somehow able to access the
secure area and remove the material without attracting attention—is even
more difficult to believe. Either way, this set the tone for what was to follow
and led to a climate of suspicion and fear, with working relationships and
friendships being stretched to the breaking point.

2. THE MORNING AFTER

In the aftermath of the UFO encounter, rumors are circulating around the base like wildfire. As the story gets passed from person to person, it evolves and takes on a life of its own. Officers and senior NCOs struggle to ensure that the enlisted men and women under their command concentrate on their normal duties. The fact that it's Christmastime doesn't help. Most of the people here would far rather be on leave or back home in the United States, with their families. It's the sort of situation that fills commanders with dread in a hierarchical, rule-based organization where things are supposed to go by the book. And that's the problem; usually, in the military, there's a Standard Operating Procedure (SOP) or protocols for every scenario that might arise. That way, nobody has to guess what to do or make up policy on the hoof. There's a set way to respond to something: scenario x is responded to by action y. Everyone learns it and then practices it until responses are almost instinctive. It's all part of the "train hard, fight easy" philosophy that's central to military life—though veterans of the campaigns in Iraq and Afghanistan smile wryly at the idea that there's much "easy" fighting these days.

So in relation to an unusual event such as a UFO sighting, where the actions to be taken might not be readily apparent, the first instinct would be to reach for the manual and see what it says. The only problem is that while there used to be a procedure for dealing with this, there isn't anymore. The US government once had an official UFO investigation program, but it was terminated in 1969.

The newly formed (it had just become a separate entity from the US Army) USAF set up a small UFO research and investigation unit in 1948, under the code name Project Sign. Unsurprisingly, this program soon acquired the nickname Project Saucer. Through a series of internal reorganizations and policy initiatives, the project name changed from Sign to Grudge in 1949. Finally, in 1952, there was another change and UFO investigations were handled under the name by which they became best known: Project Blue Book. But Project Blue Book was wound up in 1969, and to this day the only information available about UFOs on the Department of Defense (DoD) Web site is a brief history of Blue Book, followed by this statement:

> As a result of these investigations, studies, and experience, the conclusions of Project Blue Book were:
>
> No UFO reported, investigated, and evaluated by the Air Force has ever given any indication of threat to our national security.
>
> There has been no evidence submitted to or discovered by the Air Force that sightings categorized as "unidentified" represent technological developments or principles beyond the range of present-day scientific knowledge.
>
> There has been no evidence indicating that sightings categorized as "unidentified" are extraterrestrial vehicles.

Even assuming that a paper copy of this was readily available to base personnel in 1980, this would have been of no help whatsoever, because not only had the project been discontinued, but also the unit was clearly set up to deal with reports of lights or objects in the sky. The possibility that one of these UFOs might actually land seemed not to have occurred to anyone. Commanders searching for anyone in the USAF—or anywhere else in the US government—to hand this off to would soon have realized they were on their own. That only left the option of trying to hand off the problem to the Brits! Penniston was frustrated by this policy vacuum but dealt with it as best he could:

It was frustrating that neither the Air Force nor our base had an unclassified SOP for this security incident. However, with that in mind, I did improvise by using two existing ones: Aircraft Crash Security Response Option (SRO) and then when determined, the Helping Hand Security Response Option. When told by the Shift Commander during the debriefing the morning after, I felt it was understandable that there was no way to report this as a UFO incident. However, to the contrary, I do believe that transmitted reports were made from the 81st Tactical Fighter Wing (TFW) Command Post and up-channeled. This was due to the mission of the then largest Tactical Fighter Wing in the free world.

Bentwaters and Woodbridge were American bases on British soil. This is a legacy from the Second World War and the Cold War, and while the US presence in the United Kingdom has declined, there are still a number of American bases in Britain. The US presence in the United Kingdom has always been a contentious issue with the British, many of whom resent foreigners on their soil. "Overpaid, oversexed, and over here" was the wry wartime joke about Americans, which was somewhat ungenerous, given the fact that D-day (and indeed victory in the Second World War itself) would never have happened without the massive amount of men and equipment provided by the American government. After the end of the war, the resentment lingered, especially as Britain languished in a postwar austerity to the extent that rationing was not finally abolished until 1954. But again, the fact was that without US bases in the United Kingdom (and elsewhere in Europe) there would have been nothing—short of the unthinkable option of using nuclear weapons—to stop the Soviets invading Western Europe. Without American forces, estimates were that European armies would be capable of withstanding the onslaught of Soviet tanks for merely a few hours before being overrun.

UK bases made available to the United States are commanded by a US officer, usually at colonel rank. In the case of Air Force bases, despite being manned almost exclusively by USAF officers, the bases are prefixed "RAF" (Royal Air Force), maintaining a quaint fiction that these are British bases. Even more confusingly, there's an "RAF commander." Normally, this is a more junior officer, usually at squadron leader (a UK rank two levels below colonel) rank. The role of the RAF Commander is to liaise with the US Base Commander and act as head of establishment for any UK MoD employees, but to all intents and purposes the main duty is to act as the conduit between the US commander and the MoD.

The fact that there are US bases on UK soil makes for a fairly complex situation in terms of law, jurisdiction, et cetera. There are a number of documents relevant to this, including the North Atlantic Treaty, signed on April 4, 1949; the NATO Status of Forces Agreement dated June 19, 1951; and the Visiting Forces Act of 1952. These top-level pieces of legislation are supplemented by a whole range of Memoranda of Understanding, often setting out more detailed, local arrangements between individual US bases in the United Kingdom and local authorities—including the local police, though Penniston is cynical about how this worked in practice: "From the security side of things, there was no working relationship between Security and either the local civil police or the MoD Police."

The upshot of all this legislation and associated agreements was that the US Base Commander had responsibility for security of the base. In this context, security covers law enforcement and policing. There's a common misperception that US jurisdiction stops at the perimeter fence and that security "outside the wire" is the responsibility of the UK authorities, be it the civil police or, in some respects, the MoD Police. This is not correct and the actual situation is somewhat more complex. Common sense exposes the fallacy of the belief that US jurisdiction stops at the perimeter fence. Imagine a terrorist incursion that led to a firefight between attackers and US security personnel. It would clearly be nonsense to suggest that in a fluid operational situation US personnel should be unable to go beyond the fence when such a move might make tactical sense. Similarly, it would be crazy to suggest that the United States should ignore a perceived threat or another situation (such as an aircraft crash) where there's imminent danger to human life and seek to pass the responsibility to UK personnel who might be miles away and unable to respond in sufficient time. Accordingly, one finds vague phrases such as "Operations outside the bases are subject to arrangements with UK authorities" in various bilateral US/ UK agreements and other relevant documents. But the situation is confusing and all this caused difficulty, embarrassment, and a near diplomatic incident in relation to these UFO sightings.

Despite—or perhaps because of—the lack of a formal SOP on UFOs or unexplained phenomena, some personnel at the base did attempt to make some informal attempts at an investigation the day after the initial sightings. Those concerned were doubtless unaware that they were driving a coach and horses through the defensive and prickly "we don't investigate UFOs anymore" response that was given routinely by the USAF, the DoD, and even the National Aeronautics and Space Administration (NASA) when asked about the subject.

Major Edward Drury was Deputy Squadron Commander to the more senior Major Malcolm Zickler, who was in command of the 81st Security Police and Law Enforcement Squadron. Drury had been woken up and briefed by the on-duty Flight Security Officer, Lieutenant Fred "Skip" Buran, who also notified Major Zickler and Colonel Conrad, the Base Commander. These are the SOPs for a security incident and they were followed by Buran. Drury and the on-duty shift commander on the morning of December 26, Captain Mike Verrano, had been among the first two officers to question Burroughs and Penniston. Drury's first thought was that the whole affair had been a Christmas prank, but the debriefings with Burroughs and Penniston, coupled with the confirmation that the UFO had been briefly tracked on radar from the Bentwaters Command Post, soon disabused him of this theory.

At some point on the morning of December 26, Drury and Verrano decided to inspect the landing site themselves. Buran ordered Burroughs and Penniston to rendezvous with them. Also ordered to attend was Master Sergeant Ray Gulyas, whose job it would be to take measurements at the landing site, along with photographs.

Burroughs and Penniston retraced the route they had taken the night before. Arriving at the point where they'd had to dismount from their vehicle and proceed on foot, they saw nobody, so went deeper into the forest, to the landing site. Burroughs immediately saw the indentations that they'd seen in the dark and called Penniston over. Penniston paced around, measuring the distance between the three marks. Everything was exactly as they recalled it from the night before.

At this point, Drury, Verrano, and Gulyas arrived. Burroughs and

Penniston briefed them, showed them the landing site, and left. Drury, Verrano, and Gulyas stayed at the landing site for a while and then went back to brief Major Zickler. Frustratingly, Zickler instructed them to return immediately and rendezvous with one of the British police officers who had been called out the preceding night, who was concerned that he might have missed some evidence in the dark. This explains the discrepancy between the initial police report stating that nothing was found and the later report confirming that a landing site had been located.

In the meantime, Penniston had returned to his lodgings in the nearby city of Ipswich. But sleep would not come after what he had experienced and he was overtaken—as if by a compulsion—to have more evidence. He called a British friend who was an interior decorator and picked up some plaster of Paris, a jug of water, and a small bucket. Penniston then drove back to Woodbridge, put the items into his knapsack, and went back to the landing site. He mixed and poured the plaster of Paris into the three holes, waited for around an hour while the mixture set, then removed the casts, wrapped them in plastic, and placed them in the knapsack. As he was leaving, he ran into Drury, Verrano, Gulyas, and the British police officer. Drury asked Penniston what he was doing and Penniston said that he was just trying to put the events into their proper context by re-examining the landing site. Drury said they were going to handle the situation and told Penniston to go home, get some sleep, and let him worry about the investigation. Penniston departed. He said nothing about the plaster casts. This would be his secret—personal confirmation that what he and Burroughs experienced had been real: "I wanted something for me to have that was physical evidence of what had happened. I was driven harder by this thought after my meeting with the Shift Commander earlier. I had every indication this was going to fade fast from radar."

As Penniston returned to Ipswich, Drury, Verrano, Gulyas, and the British police officer proceeded to the landing site. Gulyas took his measurements and shot an entire roll of film, which he later handed to Verrano. Verrano subsequently told him that all the pictures had been fogged. In a telling foretaste of the suspicion that would soon infect many of the participants in these strange events, Gulyas returned to the site later to take his own photos. While the quality is comparatively poor, a few of the black-and-white images survived. Bizarrely, Gulyas, like Penniston, took plaster casts of the indentations on the ground—again, on his own initiative.

A few things are clear from all this. First, the policy vacuum in relation to UFOs is already causing confusion over who should be doing what. Far from being nipped in the bud by someone taking charge, this problem would linger, fester, and eventually explode into what nearly caused a diplomatic incident between the American and British governments. But already we have a sense of the problem, with UK police officers being called out during the night, departing, and then being called back the following morning, various USAF personnel wandering out to the landing site, and witnesses returning to the site of the encounter. This latter issue is the most interesting. Penniston wasn't just shell-shocked; it was as if he was under some sort of compulsion: drawn back to the landing site. Part of it was the understandable desire to get some sort of confirmation that this had actually happened—to locate some more evidence. Burroughs and Penniston had already revisited the site of their encounter for that very reason, in the dark, before returning to base on the night of December 26. The indentations and the tree damage had provided them with assurance that the incident had indeed taken place, but there may have been something deeper at work and the issue of a compulsion is a recurring theme in this whole story.

The twin concepts of looking for confirmation and searching for evidence applied not just to the witnesses but also to the chain of command, with officers such as Drury and Verrano. Again, part of this was good management (the chain of command's responsibility to the more junior ranks) and part of it was doubtless curiosity—who wouldn't be interested in taking a look at the site where military witnesses saw a UFO land? But with so many of these witnesses, it almost seems as if they were *drawn* there by something and that they were looking for something more than answers—some deeper meaning, perhaps.

It's particularly strange to note the way in which different people were doing the same thing, independently of one another. Several people took photos of the landing site the morning after the encounter. And when Gulyas returned to the site to take more photos, having his suspicions concerning the story that the first set had not come out, he—like Penniston—poured plaster of Paris into the indentations to take a mold.

While taking photos is arguably a fairly logical step, taking a mold of the indentations is a somewhat more abstract idea.

On the point about photographs, anyone who sent their film to the base's photo lab was invariably told that their film had been fogged and that no images had come out. There's controversy over whether this was part of a cover-up or was a true statement, with the fogging being caused by high levels of radioactivity at the landing site.

Just when it seemed that things were dying down a little, the unthinkable happened and the UFO returned. On the evening of December 27, a Combat Support Group awards dinner was taking place at Woody's Bar on Woodbridge. The Base Commander, Colonel Ted Conrad, was present, as was Charles Halt, the Deputy Base Commander. At some stage in the proceedings, Lieutenant Bruce Englund, the on-duty shift commander, entered the premises looking shell-shocked, took Halt aside, and told him, "It's back." Halt looked confused for a moment and asked, "What's back?" The response from Englund was clear: "The UFO is back, sir."

Halt conferred with Conrad over what to do. It was clear that one of them would have to take charge of the situation and go out into the forest. It was Conrad's decision to make, as the senior officer. Perhaps because he felt it was important for him to present the awards, perhaps because he was skeptical that there was any substance to the UFO sightings, or perhaps because, as the old military saying goes, "rank has its privileges," Conrad made his decision: he would remain at the social function. Halt would go out into the cold December night, into the forest, and investigate these latest UFO sightings. As he walked out of the building, he had no idea that he would be walking into history.

3. INTO THE DARKNESS

As Halt left the social function, his mind-set was a mixture of frustration, determination, and curiosity. It was a bitterly cold night, and as he glanced back at the lights in Woody's Bar there could have been no doubt in his mind that he'd drawn the short straw—or, rather, been handed it, by his boss. But while Halt would far rather have stayed indoors in the warmth and comfort of the social club, enjoying a convivial evening with colleagues and friends, he had a job to do. He was frustrated by the UFO rumors that were proving to be such a distraction to the men and women on the twin bases, and now, at least, had the perfect opportunity to lay the matter to rest. But he was curious, too, and who wouldn't be? After all, it's one thing to see newspaper stories about hicks seeing UFOs in the middle of nowhere or maybe catch an overly dramatic TV documentary on the subject (which was the limit of most people's exposure to the subject in those pre-Internet days), but it's quite another thing when personnel at two key military bases start seeing UFOs—and not just lights in the sky but something considerably more up close and personal. So despite his inherent skepticism and desire to bring the matter to a close, at the back of Halt's mind there was a spark of interest and curiosity. Maybe, just maybe, there was something interesting out there.

Halt's first task was to assemble a team. Lieutenant Bruce Englund, who had come to Woody's Bar to report the return of the UFO, was automatically on the team by virtue of the fact that he was the on-duty shift commander. Halt then called the Disaster Preparedness Office and spoke to the chief to see who was on standby. The individual concerned was Sergeant Monroe Nevels. Nevels was an experienced photographer and brought his camera as well as the piece of equipment more central to his duties: a Geiger counter, used to measure levels of radiation. The other member of the team was Master Sergeant Bobby Ball, the on-duty flight chief.

While Nevels had brought his Geiger counter and camera, Halt brought the handheld cassette recorder that he habitually carried with him (like Agent Cooper in the cult TV series *Twin Peaks*) to document his observations and thoughts. He also took spare batteries and several microcassettes. Other items included flashlights, radios, a starlight scope (i.e., a night-vision device), and some utility jackets, in view of the extreme cold.

In parallel, Halt ordered that the area of the forest where the UFO had been seen should be illuminated with light-alls. A light-all is essentially a set of floodlights mounted on a wheeled platform. A gasoline engine drives a generator that powers the lights. Light-alls are used to supply emergency or remote-area lighting or additional lighting for tasks such as munitions loading and aircraft maintenance. But for some reason, the light-alls were not functioning properly. There's confusion about this. Some witnesses say that the light-alls were simply low on gas, and this is certainly what the frustrated Halt believed, as he sharply ordered that they be refilled. But others involved have said that there was something odd about the malfunctioning light-alls, as if they were being interfered with in some way, perhaps akin to the way in which the radios had behaved during the initial encounter with the UFO on the first night of activity.

One consequence of the malfunctioning light-alls was the fact that more people got involved in the unfolding situation and more still heard about it. The situation was already confused, with personnel unsure whether to refill the light-alls with gasoline or fetch new ones from the motor pool. Simultaneously, through a combination of people monitoring radio traffic and simply talking to one another, word spread quickly around the twin bases that another UFO event was unfolding and that the Deputy Base Commander was investigating personally. Some of these personnel were on-duty and some were off-duty. In what sounds suspiciously like a breakdown of discipline, a number of personnel headed out into the forest without any orders or authority, simply out of curiosity. Halt was sometimes not sure who was out there as a result of his orders to provide functioning light-alls and who was out there on what amounted to a private enterprise.

Airman Tony Brisciano—on-duty at the Fuels Management Branch, and the person on the receiving end of Halt's increasingly impatient demands for gasoline and light-alls—recalls that when he arrived at the military gas pumps at Woodbridge the scene was chaotic. There were numerous light-alls mounted on the backs of pickup trucks, in various stages of being fueled. There were also a number of police cars. Brisciano said that he'd never seen so many vehicles waiting for fuel at the same time and said this was particularly noticeable because this was in the early hours of the morning. Given the busy operational role of the twin bases, plus the heavy program of exercises, this gives an indication of just how major an operation was underway.

Looking back on events, Halt is always careful to make it clear how small a team he took into the forest. Was this simply an attempt to downplay the incident so as to be consistent with the official US government line that there was no longer any official interest in UFOs or, as some conspiracy theorists allege, was it something else?

There's no definitive list of exactly who else was out there aside from the original team of Halt, Englund, Nevels, and Ball, but three individuals who were undisputedly out there were Sergeant Adrian Bustinza, Sergeant Frail, and, most intriguingly of all, one of the two key witnesses from the first night's encounter, John Burroughs.

While Halt waited for functioning light-alls to arrive, he decided to head for the landing site from the first night, to conduct his own retrospective investigation.

At this point, Halt decided to start using his mini cassette recorder. Halt recorded his thoughts and observations over the next few hours, pressing the stop button when nothing significant was taking place. Accordingly, what we have today is around eighteen minutes of dialogue, broken into what might best be described as separate chunks of action, punctuated by breaks in the tape. A complete transcript of the tape can be found as appendix E, but key extracts will be quoted in this chapter, for convenience, at the appropriate points in the text.

The recording is an extraordinary piece of evidence for a number of reasons. It is indisputably genuine, as confirmed by Halt and several of the other people whose voices can clearly be heard and identified on the tape. And over thirty years on, when memories have inevitably faded and where disagreements have emerged between some of those involved, it is immutable. Their words reach out to us across the years, and while one can question their judgment and perception, they said what they said.

Notwithstanding this, as is the case with almost every aspect of this extraordinary story, there's controversy here. In late 1999 Charles Halt told author and investigative journalist Georgina Bruni that he had four or five hours of tape that—unlike the eighteen minutes in the public domain—nobody would be allowed to hear. When she pressed him for details of what was on these tapes and why the material couldn't be made available, Halt refused to elaborate.

In the opening statement on the tape, Halt gives a rough indication of his team's location and summarizes the situation with the light-alls: "One hundred fifty feet or more from the initial, I should say suspected, impact point. Having a little difficulty, we can't get the light-all to work. . . . There seems to be some sort of mechanical problem. Let's send back and get another light-all. Meantime, we're gonna take some readings from the Geiger counter and, er, chase around the area a little bit waiting for another light-all to come back in."

The next portion of the tape deals with Halt ordering Monroe Nevels to use his Geiger counter in the clearing where John Burroughs and Jim Penniston encountered the UFO on the first night. At first Nevels finds nothing significant. There's natural background radiation everywhere and his first assessment is "just minor clicks." "OK, we're still comfortably safe here?" Halt asks. There's no reply on the tape, so either the response was given while Halt had stopped the tape or the reply was inaudible or given via a nod or a thumbs-up. In any event, it's a reasonable assumption that had the answer been "no" the men would have left the area immediately.

It will be recalled that Burroughs and Penniston had found three indentations on the ground, where it seems that the UFO had come to rest. These were the holes into which both Jim Penniston and Ray Gulyas had poured plaster of Paris, to obtain a cast of whatever it was that penetrated the ground. When plotted out, the shape formed was a near-perfect equilateral triangle. The implication was that the triangular UFO had been supported, when it landed, on legs or landing pads of some sort, similar to those of the Apollo lunar landing module. The ground was rock hard, due to the frozen December temperatures, and Halt estimated that the object must have weighed several tons to cause indentations that deep. Halt instructed Nevels to concentrate his analysis on these indentations.

The landing site is scanned carefully, with the radiation levels appearing to peak in the three indentations and in the center point of the three indentations, i.e., directly underneath the center of where the craft had landed. This area seems discolored and a discussion ensues about this.

Halt: We found a small blast—what looks like a blasted or scuffed-up area here. We're getting very positive readings. Let's see, is that near the center?

Englund: Yes, it is. That is what we would assume would be the dead center.

Shortly afterwards, there's a break in the tape. When it restarts, Halt, being a methodical man, gives an order for a more thorough assessment: "OK, why don't we do this: why don't we make a sweep—here, I've got my gloves on now—let's make a sweep out around the whole area about ten foot out, make a perimeter run around it, starting right back here at the corner, back at the same first corner where we came in, let's go right back here. . . . [Heavy breathing.] I'm gonna have to depend on you counting the clicks."

Sometime during a break in the tape at least one functioning light-all has arrived, because Halt says, ". . . then I can put the light on it."

It should be remembered that investigating the landing site was not the reason Halt had assembled a team and gone out into the forest. He'd been sent out by Colonel Conrad because of a report that the UFO had returned. But in this initial part of the tape there's no evidence to suggest that the UFO is anywhere in sight and no evidence that Halt is looking for it. Given that Englund was the one who arrived at Woody's Bar with the news that the UFO had returned, it's intriguing that there's no dialogue between the two of them about this on the tape. Clearly, for whatever reason, this discussion wasn't recorded.

Sometimes Halt's tape picked up incoming radio transmissions from personnel back at base or from personnel elsewhere in the forest who were arriving with the light-alls.

Having taken extensive Geiger counter readings from the landing site and, in particular, from the three indentations and the center point of the three holes, Halt now turned his attention to the trees that surrounded the clearing. "This looks like an abrasion on the tree," Englund observes. "There may be sap marks or something on it," Halt states. Englund summarizes the situation as he sees it: "Each one of these trees that face into the blast, what we assume is the landing site, all have an abrasion facing in the same direction, toward the center. . . ." Halt's attention is caught by a particular tree: "Never seen a pine tree that's been damaged react that fast."

They're clearly taking samples, because Nevels asks, "You got a bottle to put that in?" and Halt says, "You got a sample bottle?" before Englund says, "Put in the soil." It seems that samples were taken not just of soil but also of sap and other material from the apparently damaged sides of the trees facing the clearing. A little later when talking about samples, Halt says, "Have them cut it off, and include some of that sap. . . ."

The tree damage certainly seems consistent with what John Burroughs and Jim Penniston saw on the preceding night. However, a question to be considered is the extent to which the conclusions Halt and the others were making about all this were influenced by what Burroughs and Penniston had reported. The acid test would be this: if Halt and his men had examined the clearing for some reason, randomly, without having been aware that a UFO had been reported to have landed there, would they have noticed the indentations and tree damage and, if so, would they have thought it as noteworthy as the tape indicates they did?

At the same time as Halt was examining the landing site, one of the key players was making his way back on his own initiative. After John Burroughs went off-duty at the end of the shift during which he and Jim Penniston had their encounter, he was on a three-day break and was at home in Ipswich. Mirroring what had happened to Penniston, he, too, had trouble sleeping and felt compelled to go back to the base. When he arrived, he met up with an LE sergeant called O'Brien who greeted him jovially by saying, "Hey, it's the UFO guy!"

O'Brien then told Burroughs something that shocked him to the core. He explained that there had been a further UFO sighting on the night of December 26, involving personnel from D-Flight: Lori Buoen and the desk sergeant, John Trementozzi. Buoen had seen a fiery red/orange object descend slowly into the forest. She reported that it had been surrounded by an eerie blue/white corona. When Trementozzi and other D-Flight personnel arrived, they saw red, green, and white lights in the forest. The lights would appear at one spot, disappear, and then reappear at another point. The shift commander, Lieutenant Bonnie Tamplin, together with Master Sergeant Bobby Ball, went out to investigate. At one point the Jeep that Tamplin was driving had been struck by light beams and a blue light raced through the vehicle. All the power went off and the vehicle stalled and died. Buoen and Trementozzi had been following developments on their radio and heard Tamplin call out, "Bob, Bob, where are you? I can't see anything." Buoen recalled the fear: "She was so scared—and this was our lieutenant!"

O'Brien turned to Burroughs, no longer joking, and asked, "What the hell do you think is really going on?" Burroughs had no explanation to offer. But it was clear that there had been two separate incidents, with C-Flight and D-Flight seemingly unaware of each other's encounters. It's unclear whether the failure to brief D-Flight on what happened to members of C-Flight (consistent with SOPs, which require anything that might impact on operations to be briefed) was simply a consequence of the missing blotters or reflected a decision higher up the chain of command.

By the time Burroughs and O'Brien had this conversation, numerous personnel at the twin bases were aware of the various sightings and the other strange events. By the time Halt and his team went out into the forest, perhaps as many as fifty or sixty personnel were monitoring the situation unofficially, either by listening in to radio communications or by climbing vantage points such as the control tower at Bentwaters. This was in addition to those on-duty personnel who were involved directly in some capacity, either out with Halt or—as with the personnel refueling the light-alls—supporting Halt's investigation.

While most of the off-duty "spectators" were content to take a passive role, Burroughs was not. Having been so intimately involved on the first night, he wanted to go out to see what was happening. Partly this was a desire for confirmation (that he wasn't going crazy and that he and Penniston hadn't somehow imagined the whole thing) and maybe for some form of closure. But more than that, it was a compulsion. Burroughs felt that something was actively drawing him back out into the forest.

Burroughs hitched a ride with two friends who were heading out to the East Gate. At that point he transferred into a military Jeep and moved forward to the staging area where the light-alls were. The Halt tape captures the following exchange:

Security Control to Ball: You have Airman Burroughs and two other personnel requesting to ride 'em over [sic] on a Jeep at your location.

Ball: Tell them negative at this time. We'll tell them when they can come out here. We don't want them out here right now.

The irritation in Ball's voice can clearly be heard. It's extremely unlikely he knew that Burroughs wasn't even on-duty—a fact that would doubtless have made Ball considerably more irritated. Halt, too, was irritated, but it's unclear whether his irritation was caused by a growing sense that things were getting out of hand, with too many people milling around in the forest, whether he was concerned about contamination of evidence at the landing site, or whether the spikes in radiation levels made him think it was prudent to keep people back unless they had an absolute need to be there. While Nevels never once intimated that the peaks in radiation they were recording were dangerously high—as common sense indicates he would have done had the levels been potentially harmful—there's something almost primal about the word "radiation." Especially for a layperson, who may be unaware that there's natural background radiation everywhere, the word invariably conjures up images of post-Hiroshima or -Nagasaki radiation-sickness victims or of emaciated cancer patients who have undergone chemotherapy. It would be a basic instinct of Halt and Englund, as the senior men present, to want to avoid exposing the men and women under their command to any threat.

There was also the legal position to think of. Because there are various circumstances (e.g., a perceived security situation, which was arguably the case here) where it's quite proper for US personnel to deal with a situation off-base, jurisdiction wasn't an issue. Generally speaking, though, the UK police would have primacy (jurisdiction can be concurrent—i.e., rest with more than one person or organization—but invariably one individual or body has lead responsibility) and it might look odd for so many US military personnel to be out in the forest. There was also the presentational issue to think of. How would the story play out in the press, if the media got hold of it? These were the sorts of issues that would probably have been more in Halt's mind than in the minds of more junior personnel.

A final complicating factor would have been if Halt knew—or suspected—that some of the personnel out in the forest had been armed. Again, there's no absolute prohibition on this (nobody would expect personnel responding to a terrorist incident to do so unarmed), but the set of circumstances in which this would be permissible is extremely limited. Testimony on the question of whether anyone out in the forest that night was armed is inconclusive. The official line, of course, is that they were not. Many of those involved made a point of stating that as they left the base they left their weapons, but some testimony suggests at least some personnel were armed.

For the next few minutes Halt and the three men with him continue to examine the clearing. Nevels takes further readings with the Geiger counter and pays particular attention to the trees around the edge of the clearing—focusing on the sides of them facing the clearing, where the higher levels of radiation were detected. Halt also instructs that the starlight scope be deployed. He proceeds to look at the trees through this image intensifier and is clearly impressed by the results: "Getting a definite heat reflection off the tree," he says at one point. Then, "Three trees in the area, immediately adjacent to the site, within ten feet of the suspected landing site; we're picking up heat reflection off the trees."

Halt then notices "hot spots" on the ground. "Here, someone wanna look at the spots on the ground?" and then, "Whoops, watch you don't step . . . you're walking all over 'em. . . . OK, let's step back and not walk all over 'em"—again, it's not clear if Halt is worried about evidence being lost or contaminated or he's concerned about the health consequences if his men stepped on a potential radiation hot spot. Either way, Halt is clearly spooked by this, because the tape is punctuated with observations such as, "Hey, this is eerie," and, "This is strange."

Halt then tries to correlate the hot spots with the three indentations and, in particular, the center point formed by the three indentations. The following exchange between Halt and Nevels documents this as well as giving an insight into the potential significance of all this:

Halt: You say there is a positive aftereffect?

Nevels: Yes, there is, definitely. That's on the center spot. There is an aftereffect.

. . . . .

Nevels: What does that mean?

Englund: It means that when the lights are turned off, once we are focused in and allow time for the eyes to adjust—we are getting an indication of a heat source coming out of that center spot, as, er, which will show up on—

Halt: Heat or some form of energy. It's hardly heat at this stage of the game [i.e., nearly two days after whatever landed in the clearing departed].

Englund: And it is still . . .

There's then another break in the tape. The next recording is a monologue in which Halt describes some more tree damage: "Looking directly overhead one can see an opening in the trees, plus some freshly broken pine branches on the ground underneath. Looks like some of them came off about fifteen to twenty feet up. Some of the branches [are] about an inch or less in diameter."

Again, this would seem to be consistent with what John Burroughs and Jim Penniston reported previously. It was a small clearing and the implication is that the UFO had to maneuver its way in from directly above, which would almost inevitably have involved causing some damage to the canopy and other branches lower down. This is an assumption, of course, as Burroughs and Penniston didn't see the craft arrive. But they saw it depart and the slow ascent—prior to the "speed—impossible" departure—seemed consistent with the sort of careful maneuver required to negotiate a small hole in the canopy.

Then there's another break in the tape.

Halt has described the hot spots as being "eerie" and "strange." But the whole situation is about to get considerably more strange.

4. IT'S COMING THIS WAY

Thus far, Halt, Englund, Ball, and Nevels have gone out into the forest in response to a report that the UFO seen two nights previously and encountered at close quarters by John Burroughs and Jim Penniston has returned. Halt called for light-alls to illuminate the forest, but they're either malfunctioning, low on gas, or both. While waiting for some functioning light-alls to arrive, Halt has led the three men to the clearing where the UFO was seen to land in the early hours of December 26 and has undertaken an examination of the scene, noticing indentations on the ground and damage to trees that's attributed to the UFO. Radiation readings have been taken with a Geiger counter and seem to correlate with where the UFO was seen to land, and "hot spots" have been recorded with a night-vision device.

Halt has been recording his observations on his handheld cassette recorder. After a lengthy discussion about the "hot spots" there's another break in the tape. When the tape resumes, Halt does something that he hasn't done before and helpfully gives the time. The observation that follows is interesting: "Oh-one-forty-eight. We're hearing very strange sounds out of the farmer's barnyard animals. . . . They're very, very active, making an awful lot of noise." In fact, though the landing site is some distance into the forest, as reached from RAF Woodbridge, it's very near the boundary between the easternmost edge of the forest and a large field. The field is part of Green Farm, in the tiny hamlet of Capel St. Andrew.

There's dispute about what animals were on this farm in 1980 or in any location sufficiently close to be audible to Halt and the others. Interestingly, one thing Halt *doesn't* mention, which we know were at the farmhouse and on other nearby properties, is dogs. This suggests that it wasn't the farmyard animals that Halt heard, as this would undoubtedly have set off the dogs. It's possible that what the men heard were muntjac deer (sometimes known as barking deer), which can be found in Rendlesham Forest. Muntjac deer are nocturnal, so it may be that there was nothing unusual about these noises, which could have been cries of alarm (and it's possible that it was the presence of Halt and the others that disturbed them in the first place!) or noises associated with fighting or mating.

However, what happened next casts doubt on these mundane potential explanations and suggests—as Halt believes—that something else disturbed the animals.

Halt: You just saw a light? Where? Wait a minute. Slow down. Where?

Englund: Right on this position here. Straight ahead in between the trees—there it is again. . . . Watch—straight ahead off my flashlight, there now, sir. There it is again.

. . . . .

Halt: Hey, I see it, too. What is it?

Englund: We don't know, sir.

. . . . .

Halt: It's a strange, small red light, looks to be out maybe a quarter to a half mile, maybe further out. I'm gonna switch off.

A full transcript of the tape can be found as appendix E, but what follows are the key extracts from the next few minutes, as the UFO is observed by Halt and the others. These are in the correct order but with some less relevant quotes edited out or shortened, so as to give readers a clearer idea of what was being seen in relation to the UFO:

Halt: Is it back again?

Englund: Yes, sir.

Halt: Well, douse the flashlights then. Let's go back to the edge of the clearing so we can get a better look at it. See if you can get the star scope on it. The light's still there and all the barnyard animals have gotten quiet now.

. . . . .

Halt: . . . We're about one hundred fifty or two hundred yards from the site. . . . Everything else is just deathly calm. There is no doubt about it—there's some type of strange flashing red light ahead.

Englund: Sir, it's yellow.

Halt: I saw a yellow tinge in it, too. Weird! It appears to be maybe moving a little bit this way? It's brighter than it has been. . . . It's coming this way. It is definitely coming this way.

Unknown: Pieces of it shooting off . . .

Halt: Pieces of it are shooting off.

. . . . .

Halt: There is no doubt about it. This is weird!

Unknown: To the left!

. . . . .

Nevels: Two lights—one light just behind and one light to the left.

Halt: Keep your flashlights off. There's something very, very strange. . . .

. . . . .

Halt: OK. Pieces are falling off it again!

Englund: Sir, it just moved to the right.

. . . . .

Halt: Strange! . . . OK, we're looking at the thing; we're probably about two to three hundred yards away. It looks like an eye winking at you. Still moving from side to side, and when you put the star scope on it, it sort of like has a hollow center, a dark center; it's . . .

Englund: . . . like a pupil—

Halt: Yeah, like a pupil of an eye looking at you, winking. And the flash is so bright to the star scope that it almost burns your eye.

. . . . .

Halt: We've passed the farmer's house and are crossing the next field and we now have multiple sightings of up to five lights with a similar shape and all, but they seem to be steady now rather than a pulsating. . . . We've just crossed the creek.

Halt: At two-forty-four we're at the far side of the farmer's . . . the second farmer's field and made [a] sighting again about one hundred ten degrees. This looks like it's clear out to the coast. It's right on the horizon. Moves about a bit and flashes from time to time. Still steady or red in color. . . .

Halt: Three-oh-five. We see strange, uh, strobe-like flashes to the . . . rather sporadic, but there's definitely something . . . uh, some kind of phenomenon. . . . [D]irectly north, we've got two strange objects, ah, half-moon shape, dancing about, with colored lights on 'em. . . . The half-moons have now turned into full circles as though there was an ellipse—eclipse or something there for a minute or two.

In the next-to-last segment on the tape, we hear Halt and his team experience an extremely close encounter. Listening to the recording, one can clearly hear a mixture of bewilderment, tension, excitement, and fear in the men's voices over the next few exchanges:

Halt: Now three fifteen. Now we've got an object about ten degrees directly south. . . .

. . . . .

Nevels: To the left.

. . . . .

Halt: Ten degrees off the horizon. . . . And the ones to the north are moving—one's moving away from us.

. . . . .

Nevels: Moving out fast.

. . . . .

Halt: They're both heading north. Hey, here he comes from the south—he's coming toward us now!

Unknown: Weird.

Halt: Now we're observing what appears to be a beam coming down to the ground!

Unknown: Colors!

Halt: This is unreal. [Incredulous laugh.]

At the point when Halt observes the beam of light coming down to the ground panicked shouts can be heard in the background. Halt has clarified that what happened at this point was that a pencil-thin beam of light, like a laser, struck a point on the ground directly in front of them, about ten feet ahead, illuminating the ground. As he speculated in later years, "We just stood there in awe, you know? Is this a warning, is this a signal, is this a communication? What is this? A weapon?"

There's another break in the tape and then Halt makes the penultimate remark in this section: "Halt: Three thirty, oh-three-thirty, and the objects are still in the sky, although the one to the south looks like it's losing a little bit of altitude. We're turning around and heading back toward, uh, the base. The object to the south—the object to the south is still beaming down lights to the ground."

There's a final break, before the very last observation: "Halt: oh-four-hundred hours. One object still hovering over Woodbridge Base at about five to ten degrees off the horizon, still moving erratic and similar lights and beaming down as earlier."

It was an ironic situation. Having left Woodbridge Base several hours earlier, in response to a report that a UFO had again been seen in the forest, Halt and his men had encountered several UFOs, one of which had fired a light beam virtually at his feet. But now, as he was some distance from the base, the UFO was directly over Woodbridge, firing beams of light down at the base. It was as if the intelligence behind the craft (and Halt has never once doubted that there *was* an intelligence involved) was toying with them and showing Halt just how easy it was to penetrate the area's air defenses and operate with total impunity over one of the most sensitive military bases in the NATO alliance.

But what about John Burroughs? Halt has been oddly reluctant to acknowledge the presence of Burroughs on this third night of activity, despite the evidence from his own tape recorder. Burroughs recalls that after having initially been denied permission to join Halt and his team, he was eventually allowed to do so. He recalls that he and Sergeant Adrian Bustinza were then authorized to approach one of the lights, on the basis that Burroughs would be able to ascertain whether or not this was the same thing he and Penniston encountered on the first night. Burroughs describes what happened next:

> All of a sudden in front of us we had a blue transparent light come streaking towards us and then a white object kind of appeared up above and then floated down and was sitting out there in the distance. I asked for permission to go towards it to see if I could get a closer look. As we started going towards it, it appeared to start coming towards us. Sergeant Bustinza was on my right. He went down to the ground. He saw me go into the light. He saw me disappear. He saw the light explode and I was gone for several minutes before I reappeared. I have no recall of it. I have no memory of what happened. The next thing I know I was standing in the field and whatever it was, was gone. It was like, "What just happened?"

From conversations with Bustinza in later years, Burroughs believes that what happened was a repeat of what happened on the first night. In other words, after an "explosion of light" the triangular craft appeared from the light, but he had somehow gone "into the light," as he believes he did on both nights.

5. CHARLES HALT OVER THE YEARS

Until John Burroughs and Jim Penniston decided to speak out, Charles Halt has probably been the person most closely associated with the Rendlesham Forest incident. In the absence of the documents and photos that have mysteriously disappeared, his tape recording is one of the key pieces of evidence concerning the second encounter in the forest, on the night of December 27/28. But while a good intelligence analyst will always want to see the raw data (in this case, the original tape recording), it's no less important to summarize this material so as to give a more easily digestible account. And who better to summarize this raw data than Halt himself, so he can add some color? In that way, we can see what events strike him as important and see what he actually *thinks* about all this, particularly with the benefit of hindsight.

Fortunately, there is a way we can do this. Over the years, in a number of different places, Halt has made various statements about what happened. As is often the case, the UFO and conspiracy theory community either are not aware of this material, misinterpret it, or misrepresent it—whether these viewpoints be those of believers or skeptics. Against this background, it's important to look at this material carefully.

On November 12, 2007, Charles Halt was one of a number of panelists at an extraordinary press conference held at the National Press Club in Washington, D.C. The event was organized and sponsored by James Fox, a documentary filmmaker, and Leslie Kean, an investigative journalist who headed up an organization called the Coalition for Freedom of Information. The event featured a number of retired government, military, and aviation community personnel who had either seen a UFO while on duty or undertaken an official investigation into UFO sightings. The event was moderated by the former Arizona governor Fife Symington III.

Charles Halt was one of the panelists and after setting out the background to the sightings he described events as follows:

> We suddenly observed a bright red/orange oval object with a black center. It reminded me of an eye and appeared as though blinking. It maneuvered horizontally through the trees with occasionally vertical movement. When approached it receded and silently broke into five white lights which quickly vanished. We moved out of the forest into a pasture and observed several objects with multiple lights in the northern sky. They changed in shape from elliptical to round.
>
> Several other objects were seen to the south. One approached at high speed and sent down a concentrated beam near our feet. Another object sent down beams into the weapons storage area.

He concluded with the following assessment:

> I have no idea what we saw but do know whatever we saw was under intelligent control.

On June 17, 2010, Charles Halt signed a notarized affidavit giving a brief summary of what he saw and setting out his view of this. The document was witnessed by Katherine C. Shaw, a Virginia public notary. The document described the encounter thus:

> . . . our security team observed a light that looked like a large eye, red in color, moving through the trees. After a few minutes this object began dripping something that looked like molten metal. A short while later it broke into several smaller, white-colored objects which flew away in all directions . . .
>
> . . . someone noticed a similar object in the southern sky. It was round and, at one point, it came toward us at a very high speed. It

stopped overhead and sent down a small pencil-like beam, sort of like a laser beam. That illuminated the ground about ten feet from us and we just stood there in awe . . .

This object then moved back toward Bentwaters, and continued to send down beams of light, at one point near the Weapons Storage Area.

Halt concluded with the following assessment:

I believe the objects that I saw at close quarter were extraterrestrial in origin and that the security services of both the United States and the United Kingdom have attempted—both then and now—to subvert the significance of what occurred at Rendlesham Forest and RAF Bentwaters by the use of well-practiced methods of disinformation.

This is nothing short of sensational. Note how by 2007 Halt had discarded his careful phraseology about having "no idea" what the craft he saw was but stated that it was "under intelligent control"—a phrase that left the door open for the possibility of a secret, prototype aircraft or drone. His "I believe the objects that I saw at close quarter were extraterrestrial in origin" statement leaves no doubt whatsoever about his assessment.

It's a hypothetical point, but it's intriguing to speculate what the reaction would have been from his chain of command, the media, and the public had Halt gone public with such a statement at the time of the incident!

Halt's next point is no less incendiary. To accuse the US and UK security services of subversion and disinformation in relation to this incident is an extraordinary statement from a man whom I know to be extremely careful with every word that he says. Note, for example, how he's careful to follow the official "neither confirm nor deny" line in relation to the presence of nuclear weapons. It's an accusation that—had he made it at the time—would almost certainly have resulted in his being removed from the United Kingdom, reprimanded, and possibly dismissed from the service.

On September 22, 2012, Charles Halt was one of the panelists at an event held at the National Atomic Testing Museum in Las Vegas. The event was titled "Military UFOs: Secrets Revealed" and it was almost without precedent for a Smithsonian-affiliated institution to host an event on UFOs.

Halt gave a twenty-minute presentation that was in line with the previous two statements quoted, but—in view of the available time—fleshed out some of the details. He essentially repeated his two key points, first, that he believes the UFO he and others saw was extraterrestrial in origin and second, that elements within the US and UK intelligence community were responsible for covering up the incident. On this latter point, Halt said this: "I've heard many people say that it's time for the government to appoint an agency to investigate . . . [UFOs]. Folks, there is an agency, a very close-held, compartmentalized agency that's been investigating this for years, and there's a very active role played by many of our intelligence agencies that probably don't even know the details of what happens once they collect the data and forward it. It's kind of scary, isn't it?"

In another direct quote on the same issue, Halt stated: "I'm firmly convinced there's an agency and there is an effort to suppress."

He went on to say this: "In the last couple of years the British have released a ton of information, but has anybody ever seen what their conclusions were or heard anything about Bentwaters officially? When the documents were released, the timeframe when I was involved in the incident is missing—it's gone missing. Nothing else is missing."

Halt went on to address the question of why he hasn't suffered any adverse consequences from speaking out about this incident in such robust terms: "Probably for a couple of good reasons; number one, my rank and some of the jobs I've held; but also, very early on, I sat down and made a very detailed tape and made several copies of everything I know about it and they're secluded away. Maybe I'm paranoid. I don't know, but I think it was time well spent when I made the tapes."

Phrases such as "sat down and made a very detailed tape" and "very early on" strongly suggest Halt was *not* referring to the tape recording that he made at the time of the incident but to a more detailed tape, recorded shortly after the event, perhaps as some sort of "insurance policy." Halt has not elaborated on this point.

In the question and answer session that followed the individual presentations, Halt clashed with Colonel John Alexander. Alexander is a retired US Army officer who undertook an official search for a UFO-related group operating somewhere within government, after the termination of Project Blue Book. Alexander's group of military, intelligence community, and aerospace industry officials all had Top Secret/Sensitive Compartmentalized Information security clearances and was called the Advanced Theoretical Physics Group—they deliberately avoided using the term "UFO" so as to avoid falling within the scope of any UFO-related Freedom of Information Act (FOIA) requests. They concluded that UFOs were "real" but found no evidence of any covert group studying the phenomenon.

When Alexander told Halt there was no cover-up, Halt called him naïve and implied it was arrogant to assume that just because Alexander's group didn't find something it didn't exist. Halt went on to suggest that such a group might have been deliberately moved to the private sector (doubtless to some company where the senior figures are retired military and intelligence personnel), with the twin aims of lessening the scope for congressional scrutiny and taking it outside the scope of the Freedom of Information Act. Halt has a fair point here, and it's worth noting that while access to classified information depends to some extent on your security clearance, it also depends on your "need to know." If the "information owner" judges you have no "need to know," you won't get access, no matter how high your security clearance—or rank.

Penniston is clear who he thinks has it right: "I fully support Colonel Halt on his assessment and there was a cover-up (containment) initiated from the outset. Halt is right."

Burroughs offers this assessment: "Alexander is still following the company line about what he knows. Halt seems to be opening up with new details on what he has known for years."

While these wider issues of secrecy and access to classified information can be endlessly debated, it's important not to lose sight of the central and explosive nature of Halt's statements, namely, that the Deputy Base Commander of two of the most important bases in the NATO alliance encountered a UFO in close proximity to the installations, thinks it was extraterrestrial in origin, and believes that this was covered up by American and British intelligence.

These are about as sensational a group of UFO-related claims as a senior military officer could make.

6. THE MOST IMPORTANT BASES IN NATO

Most of the witnesses to the Rendlesham Forest incident, along with many other people who were posted to the bases at Bentwaters and Woodbridge but who are not part of the story, have described the area as being "weird," "creepy," or other variations on the same theme. Because of this and in order to place these extraordinary events into their proper historical and geopolitical context, it seems appropriate to go into some more details about the area in 1980, the history of the twin bases, their role at the time of the incident, and the command structure that was in operation.

The county of Suffolk, in the United Kingdom, lies less than one hundred miles from London. But in terms of its character, it might as well be half a world away. The flat, low-lying, and largely rural landscape consists of a mixture of farmland, wetlands, and small towns. The area is rich in history, with archaeological finds and sites dating back to the Stone Age. Sutton Hoo, near Woodbridge, is the site of one of the most important archaeological sites in the United Kingdom, where two massive Anglo-Saxon burial sites dating from the early seventh century yielded a wealth of finds, including a warrior's helmet, weapons, and an entire ship!

From as far back as when the United Kingdom was first settled, before the changing geography turned Britain into an island, Suffolk was inhabited. The English Channel may have made things more difficult, but it certainly didn't stop waves of invaders and migrants coming to Suffolk to pillage, settle, or conquer. The Romans, the Anglo-Saxons, and the Vikings all left their mark on this ancient land. Pagans and then Christians wor-shiped here. The area is rich in legend and folklore, with stories of ghosts, witches, and monsters. Stories include that of Black Shuck, a ghostly black dog (often associated with the devil) said to roam the area. More recently, East End Charlie is the name given to the ghost of a German Luftwaffe pi-lot supposedly killed in the area during the Second World War. Rumors of witchcraft persist to this day, and in this sparsely populated area with its insular people, it's not hard to believe such stories.

From the early twentieth century Suffolk has played host to some of the most secretive and groundbreaking scientific and military research and development undertaken in the United Kingdom. Even now, not all the details of this can be made public. From secret sites such as Bawdsey Manor and Orfordness work was done that was to change the world.

In 1915 the Armament Experimental Squadron was based at Orford-ness, where increasingly powerful bombs were tested, against the back-drop of the First World War. In the thirties, as war with Nazi Germany approached, the Air Ministry asked Scottish meteorologist Robert Watson-Watt whether it was possible that radio waves might be focused into a beam powerful enough to incapacitate a pilot or even to destroy an aircraft. So far as we know, no such death ray was ever developed, but Watson-Watt reported that radio waves might have an alternative use, detecting incoming aircraft. A research team was set up at Orfordness and they later relocated to Bawdsey Manor, which was renamed Bawdsey Research Sta-tion. The result of this, of course, was radar and the construction of a chain of radar stations that became operational just in time to play a deci-sive role in the Battle of Britain and thus, arguably, in the outcome of the Second World War itself.

After the war, the United Kingdom's fledgling nuclear program was based at Orfordness, where the Atomic Weapons Research Establishment was based until it transferred to Aldermaston in 1971. There is also some evidence that weather modification experiments may have taken place at this site, and it may be no coincidence that the area has been hit by ex-treme weather many times over the years, from the North Sea Flood of 1953 to the Great Storm of 1987—which hit Rendlesham Forest particu-larly hard, all but flattening large swathes of the forest.

Bawdsey later became home to a secret US research project code-named Cobra Mist, designed at developing an over-the-horizon radar system.

Martlesham Heath is another interesting local site. Opened in 1917, when it housed the Aeroplane Experimental Unit, the site later became home to a British Telecommunications research facility that has close links with the United Kingdom's three intelligence agencies; MI5, MI6, and GCHQ.

The nearby village of Sizewell, on the coast, is the site of two massive nuclear power stations, Magnox Sizewell A and Pressurized Water Reactor Sizewell B.

Part of the job of a good intelligence analyst is to look for pieces of apparently separate information that, when linked, form a single, coherent picture. It's a bit like assembling a jigsaw puzzle, with the picture only becoming clear when enough small pieces (that themselves may appear unremarkable) have been assembled. Being able to spot connections is a key skill. However, humans often see patterns where none exist. An example is pareidolia—where people "see" the face of Jesus in a Danish pastry. So it is in intelligence, where looking for connections is an important skill but where it's equally important to avoid seeing connections that aren't there. It gets even more complicated when, because of an analyst's personal view or a perception of what his or her political bosses want, a conclusion-led approach is taken. A good example would be the way in which certain people went looking for a connection between the 9/11 terror attacks and Saddam Hussein's regime, because that's what they expected (and in some cases wanted) to find. Suffice to say, we highlight the history of the area not to allege or imply any connection with the Rendlesham Forest incident but to give some local "color," so readers get a better sense of the stage on which these events played out.

The twin bases of Bentwaters and Woodbridge date back to the Second World War. Woodbridge was completed in 1943 and Bentwaters in 1944. By this time, the tide of the war had turned, and a key role in this was the part played by the RAF's Bomber Command and by the US Army Air Corps (the USAF had yet to be formed as a separate military branch). Flying from bases in the United Kingdom and sometimes mounting raids where over a thousand aircraft took part, RAF and US aircraft bombed the cities and industrial sites of Nazi Germany relentlessly, pegging back

Germany's military production. But the price was high. Casualties were horrendous, and the lumbering bombers were vulnerable to German fighters and anti-aircraft guns alike. Many aircraft made the return leg badly damaged, pilots desperately using all their skills to coax their planes back over the sea to England. In such a situation, building new airfields on the east coast was a sensible tactic, as damaged aircraft were often both unstable in the extreme and desperately short of fuel. Airfields such as Bentwaters and Woodbridge were designed specifically for such damaged aircraft, to give aircrews who did make it back over the sea a location where they could quickly land damaged aircraft. The runways were among the widest and longest in the country and these "emergency airstrips" undoubtedly saved many lives.

After the end of the war, the future of the bases was uncertain, but the Second World War was soon followed by the Cold War, and as Churchill's "iron curtain" descended, cutting Europe in two, the USAF (formed in 1947) took over the twin bases in 1951.

It's important to understand the politico-military situation at the time of the Rendlesham Forest incident. With our current pre-occupation with small regional conflicts and with terrorism, it's easy to forget that in 1980 there was a very real sense that nuclear war between America and the Soviet Union—or, technically, between NATO and the Warsaw Pact nations—might break out. If we zero in on 1980 we find ourselves at a point where a developing situation would soon evolve into the greatest period of East/West tension since the Cuban Missile Crisis in 1962, when the world stood on the brink of nuclear war.

In the United Kingdom, the uncompromising Conservative Margaret Thatcher was Prime Minister, having won the General Election in 1979 to become the United Kingdom's first female prime minister. Though his inauguration was not until January 1981, Ronald Reagan had won the US election on November 4, 1980, and though the Thatcher/Reagan partnership was yet to take shape, it was clear that in both the United States and the United Kingdom a tougher stance would be taken on the Soviet Union than had previously been the case.

As it turned out, the resolve of these two hard-line politicians was soon to be tested. The seeds of the crisis they would face were sown in the unlikely location of a Polish shipyard in Gdansk, by a union official named

Lech Walesa. The Solidarity trade union had been formed on September 17, 1980, after an earlier wave of strikes. It was the first union in the Warsaw Pact not to be controlled by the Communist Party of the nation in which it was based.

It is still unclear just how close we came to World War Three at this time, not least because so much of the information is still highly classified—especially information regarding Cold War spy Colonel Ryszard Kukliński, who was spirited out of Poland by the CIA shortly before martial law was declared in December 1981. There's some intriguing historical evidence to suggest that a Soviet invasion of Poland had been scheduled for December 8, 1980, under the cover of a scheduled military exercise called Soyuz 1980. Supposedly, the invasion was only called off on December 5 when the Polish government assured the Soviets they'd deal with Solidarity internally—an assurance Soviet leader Leonid Brezhnev accepted, subject to conditions. While some historians believe the invasion threat was a Soviet bluff, designed to persuade the Polish government to launch its own crackdown on Solidarity, others are not sure, given the erratic nature of the aging Brezhnev and other Soviet leaders. Even if it was a bluff on this occasion, there was a strong possibility the Soviets would take a hard line if the Polish government couldn't control the situation. While the separate and better-known crisis that was to evolve from the January 15, 1981, meeting between Lech Walesa and Pope John Paul II was still in the (not too distant) future when the Rendlesham Forest incident took place, the Soyuz 1980 affair may have been by far the greater crisis—and may have come to a head just before the Rendlesham Forest incident took place.

Those people serving on the bases were unaware of these high-level dramas but clearly knew something was going on. Burroughs describes the general atmosphere thus: "Things were calm when I first got there, but after the mission came on line with the A-10 and Reagan became President, things became very edgy."

Even now, over thirty years after the Rendlesham Forest incident (with Bentwaters closed and operating as a commercial business park and with Woodbridge returned to the MoD and largely used as a military barracks), the full story of the twin bases cannot be told. We can give an outline of the role of the establishments, the mission, the equipment, and the personnel, but some details are still classified and must remain so. This has placed us in a difficult position. One the one hand, it is frustrating to have to leave out some key parts of the story, especially when some of them are told (with varying degrees of accuracy) by others and are freely available on the Internet. However, in other respects, there was never any choice about this. John Burroughs, Jim Penniston, and I have all taken security oaths that bind us for life, and we take this seriously, both in a legal sense and in the sense that we remain loyal to our former government masters. Divulging classified information without proper authorization is not only a criminal offence that carries severe penalties; it's also a betrayal of trust.

Burroughs and Penniston are loyal ex-military personnel who served with dedication and distinction. They have risked their lives for their country and for the ideals of freedom and democracy that they cherish. Despite the frustration they feel at having to leave out parts of the story that some people might consider important, this isn't negotiable. Values such as integrity, honesty, and loyalty are hardwired into people such as Burroughs and Penniston.

Regrettably, it doesn't always work the other way around. Despite the fact that loyalty should be a two-way street, Burroughs, Penniston, and many of the other young men and women caught up in these events feel betrayed by the chain of command. They suspect that either the chain of command (or some people therein) knew something about these events or, conversely, they ignored them because of the knee-jerk prejudice often elicited by even the use of the term "UFO." Worse still, they believe that afterwards the events were either downplayed or actively covered up and that some witnesses were threatened and drugged or, at the less serious end of the spectrum, left to get on with their careers and their lives with no counseling or other form of aftercare.

Though nominally two separate military establishments (lying close together, separated in part by Rendlesham Forest), Bentwaters and Wood-bridge operated as a single entity, the so-called twin bases having been treated in this way since 1958. They were part of a large number of USAF bases in the United Kingdom (collectively known as 3rd Air Force), coming under the control of General Robert W. Bazley, who was based at RAF Mildenhall. These USAF bases in the United Kingdom were part of

USAFE—United States Air Forces in Europe. USAFE had its headquarters in Ramstein, Germany, and came under the control of General Charles A. Gabriel.

The twin bases were home to the 81st Tactical Fighter Wing, which in 1980 operated the A-10 Thunderbolt II aircraft, nicknamed the Warthog.

The A-10 is a close air-support aircraft and its primary role is to attack tanks, armored vehicles, artillery pieces, and other high-value military targets on the ground. In the event of a conventional war between the Warsaw Pact nations and NATO, the thinking was that Western forces would be faced with a massive invasion of armor across the plains of Europe. This is where an aircraft like the A-10 would have come into its own. A significant portion of the A-10s based at Bentwaters/Woodbridge (by December 1980 an expansion to six squadrons was nearly complete) would doubtless have been deployed to Forward Operating Bases and thrown into the fray in a desperate attempt to destroy as much Warsaw Pact armor as possible, blunting an attack and thus trying to redress the balance in tanks and armored vehicles, where the numerical advantage lay with the Warsaw Pact countries.

The A-10's two main weapons systems are AGM-65 Maverick surface-to-air missiles and the 30mm GAU-8A Avenger cannon, which as operated in 1980 could fire at two rates, of either around two thousand or four thousand rounds per minute. The A-10 first saw combat in the Gulf War in 1990/91, where it's credited with having destroyed over nine hundred tanks, around two thousand other military vehicles, and twelve hundred artillery pieces. A-10s have seen subsequent action in the Balkans and in the wars in Iraq and Afghanistan.

In 1980 the twin bases were commanded by Colonel Gordon Williams. He was known as the Wing Commander, in the sense that he commanded the 81st Tactical Fighter Wing. This proved to be confusing for the MoD, because wing commander is a rank in the RAF equivalent to a lieutenant commander in the USAF. His deputy was known as the vice wing commander. Under these two commanders came four major departments: Operations, Maintenance, Rescue Management, and the Combat Support Group. Most of the personnel involved in this story were from the Combat Support Group, commanded by Colonel Ted Conrad and his deputy, Lieutenant Colonel Charles Halt. To further confuse matters, these two posts were generally referred to as the Base Commander and the Deputy Base Commander, incorrectly implying (to those unfamiliar with USAF terminology) that they were the two most senior officers at the twin bases. The reason for the Base Commander and Deputy Commander descriptors was because Combat Support Group responsibilities included the general base management duties.

It's important to clarify something about the role of the RAF. The titles "RAF Bentwaters" and "RAF Woodbridge" were misnomers—a little piece of fiction that might have glossed over matters with the public but fooled nobody else. In everything but name, these were American bases. The fact that they were on British soil was a minor detail and served only to reinforce what all military planners knew: that in any conventional war, Europe—including the United Kingdom—was utterly dependent upon US military might to defend against an invasion from the forces of the Soviet Union and the Warsaw Pact. The United States was the dominant power in NATO, just as the Soviet Union dominated the Warsaw Pact nations.

As far as the twin bases were concerned, just about the only RAF presence was a liaison officer who acted as the interface between the bases and MoD. He held the rank of squadron leader and, despite the misleading title of "RAF Commander," commanded very little. The role was an important one, but he was essentially a conduit between the USAF and his RAF/MoD bosses.

The legal status of US bases in the United Kingdom was defined in an overarching document called the NATO Status of Forces Agreement. If ever an issue arose about a question such as jurisdiction, the answer could be found "in the NATO SOFA": a phrase that always raised a few eyebrows with people on the basis that it conjured up images of looking for small change down the back of the living room furniture.

As mentioned earlier, most of the personnel involved in this story were part of the Combat Support Group. Further, most were part of the 81st Security Police and Law Enforcement Squadron. At the time of the incidents, Major Malcolm Zickler was in command of this unit. He reported to Charles Halt.

To put the duties of the 81st Security Police and Law Enforcement Squadron into everyday language, they were responsible for security and policing. In other words, the security of the base, the equipment, and the personnel was their responsibility, as was enforcing military law and discipline. Part of this involves guarding, but this is confusing, because other junior personnel from all units can (and do) find themselves on guard duty. But so far as the 81st Security Police and Law Enforcement Squadron was concerned, guarding meant having particular responsibility for high-value assets and taking a wider strategic responsibility for securing the bases. If this point seems a little labored, it's an important one, because the Security Police (SP) and LE personnel caught up in these events were highly trained personnel with a role and responsibility that went significantly further than simply patrolling the perimeter fence—important though such a task is.

While I have heard some former USAF personnel speak disparagingly about their time at the twin bases and have heard them describe the local area using words such as "creepy," "sinister," and "depressing," it is clear that Penniston and Burroughs take pride in their service and, notwithstanding the way they were treated over the UFO encounter, never lost sight of the importance of the mission. Penniston sums up the situation like this:

> I felt that the mission of the 81st TFW was a key part of deterring the Warsaw Pact nations from moving in 5000 tanks and invading what was then the Federal Republic of Germany. It was a strategic non-nuclear option which was available to the United States and NATO. The state of readiness for the 81st TFW was superb—readiness to support any aggression, when combined with our four forward operating locations in Germany and communication support from the rest of NATO. I consider my assignment to RAF Bentwaters as the best I had in my twenty years in the USAF. My fellow airmen thought the same. We all still talk often about that.

We now come to a difficult point in this book. The question arises as to whether or not there were nuclear weapons at the twin bases. Some of those who were based there at the time have said quite openly that there were. Other people who would have been in a position to know have also made statements to this effect. A good example of this is Lord Peter Hill-Norton, a former UK Chief of the Defence Staff (a UK post broadly equivalent to that of the Chairman of the Joint Chiefs of Staff) and chair of NATO's Military Committee. Writing about the Rendlesham Forest incident, Lord Hill-Norton said this:

> My position both privately and publicly expressed over the last dozen years or more, is that there are only two possibilities, either:
> a. An intrusion into our Air Space and a landing by unidentified craft took place at Rendlesham, as described.
> Or
> b. The Deputy Commander of an operational, nuclear armed, US Air Force Base in England, and a large number of his enlisted men, were either hallucinating or lying.

This quote was taken from a letter dated October 22, 1997, that Lord Hill-Norton had sent to Lord Gilbert, Minister of State at the MoD. The statement about nuclear weapons could not be clearer.

Notwithstanding the above, so far as the UK government is concerned, the position is not to comment on such nuclear questions. The following statement is a typical one and is taken from *Hansard,* the official record of proceedings of the United Kingdom's Parliament: "It is the long-standing policy of successive Governments to neither confirm nor deny the presence of nuclear weapons at any particular place or time." The final part of the preceding statement makes it clear that this "NCND" (neither confirm nor deny) policy applies retrospectively and that this would be the official response to what many might now regard as a historical query about a bygone age.

Here is a similar NCND statement from *Hansard,* which makes it clear that the policy applies equally to any query about the US government: "It is the long-standing policy of Her Majesty's Government and of the United States Government neither to confirm nor deny the presence of nuclear weapons in ships, aircraft or any particular location."

Lord Hill-Norton was well aware of the NCND policy but chose to ignore it. In October 1997 he asked the following, as a formal PQ (Parliamentary Question): "Whether the allegations contained in the recently published book Left at East Gate, to the effect that nuclear weapons were stored at RAF Bentwaters and RAF Woodbridge in violation of UK/US treaty obligations are true."

The answer, printed in the October 28 *Hansard* and signed off by Lord Gilbert, was as follows: "It has always been the policy of this and previous governments neither to confirm nor to deny where nuclear weapons are located either in the UK or elsewhere, in the past or at the present time. Such information would be withheld under exemption 1 of the Code of Practice on Access to Government Information."

There was a second question, inspired by what Lord Hill-Norton had learned about Charles Halt's UFO sighting and in particular by the final remarks on the tape recording, about the UFO firing light beams onto the base: "Whether they are aware of reports from the United States Air Force personnel that nuclear weapons stored in the Weapons Storage Area at RAF Woodbridge were struck by light beams fired from an unidentified craft seen over the base in the period 25–30 December 1980, and if so, what action was subsequently taken."

The reply was as follows: "There is no evidence to suggest that the Ministry of Defence received any such reports." The response is an interesting one, in that it doesn't simply say "no." There's an art to drafting replies to PQs, because misleading Parliament even inadvertently is taken extremely seriously, so answers have to be precise yet leave the door open in certain circumstances, e.g., when you can't be sure of the position. Playing devil's advocate, the answer to Lord Hill-Norton's second question would still be truthful and accurate in a scenario where the MoD was aware that the incident had taken place but where there was no surviving paper trail. This may sound pedantic, but there's a critical difference. When one is responding to a question along the lines of "Did x happen?" the response "We are not aware that x happened," or, "There is no evidence that x happened," or, "We can find no records that would indicate that x happened" is far safer than "No, x did not happen."

Even bearing in mind the warning about seeing connections where none exist, it's clear from all of this that the Rendlesham Forest incident took place at an extremely important location, at a particularly sensitive time. And with that observation in mind, we'll return to the aftermath of the sightings and see how the shell-shocked witnesses were treated. Because if they thought their ordeal was over, they were sadly mistaken.

7. DEBRIEFING THE WITNESSES

Various men and women at RAF Bentwaters and Woodbridge—including the Deputy Base Commander, Charles Halt—encountered something truly extraordinary. Although there's a debate to be had about what happened to them, there's no doubt that they experienced something highly unusual. Common sense tells us that this must have had a profound effect on those concerned. More than that, the reactions and behavior of some of those concerned—most notably John Burroughs and Jim Penniston, whose actions after the events border upon compulsive—clearly indicate that they were disturbed by what they saw and what they experienced. This is a natural reaction when someone is confronted by something outside their frame of reference. They have no way to process the information. This causes stress that manifests itself in a number of different ways, which can vary depending upon how closely the person was affected and depending upon their individual character and temperament. Nowadays we would probably lump all of this together under the heading Post-Traumatic Stress Disorder.

The National Institutes of Health fact sheet on Post-Traumatic Stress Disorder begins by stating:

> Post-traumatic stress disorder (PTSD) is an anxiety disorder that
> some people develop after seeing or living through an event that caused
> or threatened serious harm or death. According to the 2005 National
> Comorbidity Survey-Replication study, PTSD affects about 7.7 million

American adults in a given year, though the disorder can develop at
any age, including childhood. Symptoms include strong and unwanted
memories of the event, bad dreams, emotional numbness, intense guilt
or worry, angry outbursts, feeling "on edge," and avoiding thoughts
and situations that are reminders of the trauma.

While the reference to "an event that caused or threatened serious harm
or death" might at first glance seem a little extreme for what happened in
Rendlesham Forest, it should be recalled what Halt said about the moment
that the UFO fired a beam of light down at him and his men: "We just
stood there in awe, you know? Is this a warning, is this a signal, is this a
communication? What is this? A weapon?" However, other definitions of
PTSD allow a less extreme cause. The Merriam-Webster dictionary defines
it as a psychological condition that can occur "after experiencing a highly
stressing event"—a more subjective trigger (and with a lower threshold)
that would certainly seem applicable to the Rendlesham Forest incident.

It's ironic that it was in 1980 that PTSD was recognized as a disorder
with specific symptoms that could be reliably diagnosed and was added to
the American Psychiatric Association's *Diagnostic and Statistical Manual
of Mental Disorders*. The military, however, was slow off the mark when it
came to recognizing PTSD. Arguably, they always have been. Whether it's
the scandal of "shell-shocked" youths being executed for cowardice by
the British Army in the First World War or the tragedy of the way in
which Vietnam veterans were treated (and indeed viewed by US society as
a whole), the US and the UK military do not have a proud record on this
issue. To be fair, the US DoD and the Department of Veterans Affairs have
made substantial progress on these issues in recent years. The same is true
of the UK MoD and the Veterans Agency. But this progress, forged in Iraq
and Afghanistan and part of the post-9/11 world in which military personnel are finally enjoying the societal recognition that they deserve, is all
too recent. The situation in 1980 was very different, and even now, when
we come to consider the lobbying that John Burroughs and Jim Penniston
have done, there's no official recognition that the Rendlesham Forest incident caused any health problems, be they physical or mental. Anyone wandering around in a state of shock in the aftermath of the Rendlesham
Forest incident would have received nothing more than a brief "pull yourself

together." Weakness was—and still is—despised in the military, and even if those concerned were suffering, they knew there was no real alternative to knuckling down and getting on with their jobs.

If that had been as far as things went, one might, perhaps, accuse the chain of command of insensitivity. Nowadays one might say that the "duty of care" had not been honored or the Military Covenant (a UK concept defining the debt owed by the nation to its Armed Forces) had been breached. But while the chain of command provided no support to the Rendlesham Forest witnesses, that's not to say they were left alone.

There is, as can clearly be demonstrated, an official and an unofficial version of the debriefing process that followed the Rendlesham Forest incident. The official version is easy to document, though the problems with this documentation will be glaringly obvious.

Out of all the various witnesses on the two key nights of activity, official witness statements are available from just five individuals. In order of rank, with the most senior first, they are as follows:

Lieutenant Fred Buran
Master Sergeant J. D. Chandler
Staff Sergeant Jim Penniston
Airman First Class John Burroughs
Airman First Class Ed Cabansag

As there are so few USAF documents on the case publicly available, and because these documents were drawn up so soon after the events concerned, it's important to reproduce them in full. In all cases we've kept the spacing as in the original statements (in fact, only Buran's statement was paragraphed) but have corrected spelling mistakes. To preserve the "flavor" of these statements we've left grammatical errors as they are, except where doing so would render the material unintelligible.

LIEUTENANT FRED BURAN

This statement was typed on Air Force Form 1169 (Statement of Witness) and began with the standard phrase "I do hereby voluntarily and of my own free will make the following statement without having been subjected to any coercion, unlawful influence or unlawful inducement." It was dated January 2, 1981, and was signed at the end.

The following statement is general in nature and may be inaccurate in some instances due to the time-lapse involved and the fact I was not taking notes at the time of the occurrence. At approximately 03:00 hrs, 26 December 1980, I was on duty at bldg. 679, Central Security Control, when I was notified that A1C Burroughs had sighted some strange lights in the wooded area east of the runway at RAF Woodbridge.

Shortly after the initial report A1C Burroughs was joined by SSgt Jim Penniston and his rider, AMN Cabansag. SSgt Penniston also reported the strange lights. I directed SSgt Coffey, the on duty Security Controller, to attempt to ascertain from SSgt Penniston whether or not the lights could be marker lights of some kind, to which SSgt Penniston said that he had never seen lights of this color or nature in the area before. He described them as red, blue, white and orange.

SSgt Penniston requested permission to investigate. After he had been joined by the Security Flight Chief, MSgt Chandler, and turned his weapon over to him, I directed them to go ahead. SSgt Penniston had previously informed me that the lights appeared to be no further than 100 yds from the road East Gate of the runway.

I monitored their progress (Penniston, Burroughs and Cabansag) as they entered the wooded area. They appeared to get very close to the lights, and at one point SSgt Penniston stated that it was a definite mechanical object. Due to the colors they reported I alerted them to the fact that they may have been approaching a light aircraft crash scene. I directed SSgt Coffey to check with the tower to see if they could throw some light on the subject. They could not help.

SSgt Penniston reported getting near the "object" and then all of a sudden said they had gone past it and were looking at a marker beacon that was in the same direction as the other lights. I asked him if he could have been mistaken, to which Penniston replied that had I seen the other lights I would know the difference. SSgt Penniston seemed somewhat agitated at this point.

They continued to look further, to no avail. At approximately 3:43 hrs, I terminated the investigation and ordered all units back to their normal duties.

I directed SSgt Penniston to take notes of the incident when he came in that morning. After talking with him face-to-face concerning the incident, I am convinced that he saw something out of the realm of explanation for him at that time. I would like to state at this time that SSgt Penniston is a totally reliable and mature individual. He was not overly excited, nor do I think he is subject to overreaction or misinterpretation of circumstances. Later that morning, after conversing with CPT Mike Verrano, the day-shift commander, I discovered that there had been several other sightings. Any further developments I have no direct knowledge of.

MASTER SERGEANT J. D. CHANDLER

This statement was typed on Air Force Form 1169 (Statement of Witness) and began with the standard phrase "I do hereby voluntarily and of my own free will make the following statement without having been subjected to any coercion, unlawful influence or unlawful inducement." It was dated January 2, 1981, and was signed at the end:

At approximately 3.00hrs, 26 December 1980, while conducting security checks on RAF Bentwaters, I monitored a radio transmission from A1C Burroughs, Law Enforcement patrol at RAF Woodbridge, stating that he was observing strange lights in the wooded area just beyond the access road, leading from the East Gate at RAF Woodbridge. SSgt Penniston, Security Supervisor, was contacted and directed to contact Burroughs at the East Gate. Upon arrival, SSgt Penniston immediately notified CSC that he too was observing these lights and requested to make a closer observation. After several minutes, Penniston requested my presence. I departed RAF Bentwaters through Butley Gate for RAF Woodbridge. When I arrived, SSgt Penniston, A1C Burroughs and Amn Cabansag had entered the wooded area just beyond the clearing at the access road. We set up radio relay between SSgt Penniston, myself and CSC. On one occasion Penniston relayed that he was close enough to the object to determine it was definitely a mechanical object. He stated that he was within 50 metres. He also stated that there was lots of noises in the area which seemed to be animals running around. Each time Penniston gave me the indication that he was about to reach the area where the lights were he would give an extended estimated location. He eventually arrived at a "beacon light," however, he stated that this was not the light or lights he had originally observed. He was instructed to return. While en route out of the area he reported seeing lights again almost in direct pass where they had passed earlier. Shortly after this, they reported that the lights were no longer visible. SSgt Penniston returned to RAF Woodbridge. After talking to the three of them, I am sure that they had observed something unusual. At no time did I observe anything from the time I arrived at RAF Woodbridge.

STAFF SERGEANT JIM PENNISTON

This statement is typed on a sheet of paper headed "Statement." It is unsigned and undated and has various sketches attached. Penniston says this is the statement that was given to him at the Air Force Office of Special Investigations (AFOSI) building and was the story that he was ordered to tell anyone who asked about the event, due to the fact that there was an ongoing investigation at the time, by an "outside investigative department":

Received dispatch from CSC to rendezvous with Police 4 AIC Burroughs, and Police 5 SSgt Steffens at east gate Woodbridge. Upon arriving at east gate directly to the east about 1½ miles in a large wooded area. A large yellow glowing light was emitting above the trees. In the centre of the lighted area directly in the centre ground level, there was red light blinking on and off 5 to 10 second intervals. And a blue light that was being for the most part steady. After receiving permission from CSC, we proceeded off base past east gate, down an old logging road. Left vehicle, proceeded on foot. Burroughs and I were approx. 15–20 meters apart and proceeding on a true east direction from logging road. The area in front of us was lighting up a 30 meter area. When we got within a 50 meter distance, the object was producing red and blue light. The blue light was steady and projecting under the object. It was up the area directly extending a meter or two out. At this point of positive identification I relayed to CSC, SSgt Coffey. Positive sighting of the object . . . 1 . . . color of lights and that it was definitely mechanical in nature. This is the closest point that I was near the object at any point. We then proceeded after it. It moved in a zig-zagging manner back through the woods, then lost sight of it. On the way back we encountered a blue streaking light to the left only lasting a few seconds. After a 45 min walk arrived at our vehicle.

AIRMAN FIRST CLASS JOHN BURROUGHS

This statement is handwritten on an unheaded sheet of paper and has a sketch attached. It is signed, but not dated:

On the night of 25–26 Dec at around 3:00, while on patrol down at East Gate, myself and my partner saw lights coming from the woods due east of the gate. The lights were red and blue, the red one above the blue one, and they were flashing on and off. Because I've never seen anything like that coming from the woods before we decided to drive down and see what it was. We went down east-gate road and took a right at the stop sign and drove about 10–20 yards to where there is a road that goes into the forest. I could see a white light shining into the trees and I could still see the red and blue one. We decided we better go call it in so we went back up towards East Gate and called it in. The whole time I could see the lights and the white light was almost at the edge of the road and the blue and red lights were still out in the woods. A security unit was sent down to the gate and when they got there they could see it too. We asked permission to go and see what it was. We took the truck down the road that leads into the forest. As we went down the east-gate road and the road that leads into the forest, the lights were moving back and they appeared to stop in a bunch of trees. We stopped the truck where the road stopped and went on foot. We crossed a small open field that led into where the lights were coming from, and as we were coming into the trees there were strange noises, like a woman screaming. Also the woods lit up and you could hear the farm animals making a lot of noises, and there was a lot of movement in the woods. All three of us hit the ground and whatever it was started moving back towards the open field and after a minute or two we got up and moved into the trees and the lights moved out into the open field. We got up to a fence that separated the trees from the open field. You could see the lights down by a farmer's house. We climbed over the fence and started walking toward the red and blue lights and they just disappeared. Once we reached the farmer's house we could see a beacon going around, so we went toward it. We followed it for about 2 miles before we could see it was coming from a lighthouse. We had just passed a creek and were told to come back when we saw a blue light to our left in the trees. It was only there for a minute and just streaked away. After that we didn't see anything and returned to the truck.

AIRMAN FIRST CLASS ED CABANSAG

This statement is typed on an unheaded sheet of paper and is signed but not dated. Airman Cabansag openly admits this was a retyped statement and that he was told to sign it without question and his involvement would be over. He says this was done under extreme duress:

On 26 Dec 80, SSgt Penniston and I were on Security #6 at Wood-bridge Base. I was the member. We were patrolling Delta NAPA when we received a call over the radio. It stated that Police #4 had seen some strange lights out past the East Gate and we were to respond. SSgt Penniston and I left Delta NAPA, heading for the East Gate code two. When we got there SSgt Steffens and A1C Burroughs were on patrol. They told us they had seen some funny lights out in the woods. We notified CSC and we asked permission to investigate further. They gave us the go-ahead. We left our weapons with SSgt Steffens who remained at the gate. Thus the three of us went out to investigate. We stopped the Security Police vehicle about 100 meters from the gate. Due to the terrain we had to on by foot. We kept in constant contact with CSC. While we walked, each one of us would see the lights. Blue, red, white, and yellow. The beacon light turned out to be the yellow light. We would see them periodically, but not in a specific pattern. As we approached, the lights would seem to be at the edge of the forest. We were about 100 meters from the edge of the forest when I saw a quick movement, it look visible for a moment. It looked like it spun left a quarter of a turn, then it was gone. I advised SSgt Penniston and A1C Burroughs. We advised CSC and proceeded in extreme caution. When we got about 75–50 meters, MSgt Chandler/ Flight Chief, was on the scene. CSC was not reading our transmissions very well, so we used MSgt Chandler as a go-between. He remained back at out vehicle. As we entered the forest, the blue and red lights were not visible anymore. Only the beacon light was still blinking. We figured the lights were coming from past the forest, since nothing was visible when we passed through the woody forest. We would see a glowing near the beacon light, but as we got closer we found it to be a lit up farm house. After we had passed through the forest, we thought it had to be an aircraft accident. So did CSC as well. But we ran and walked a good 2 miles past out the vehicle, until we got to a vantage point where we could determine that what we were chasing was only a beacon light off in the distance. Our route through the forest and field was a direct one, straight towards the light. We informed CSC that the light beacon was farther than we thought, so CSC terminated our Investigation. A1C Burroughs and I took a road, while SSgt Penniston walked straight back from where we came. A1C Burroughs saw the light again, this time it was coming from the left of us, as we were walking back to our patrol vehicle. We got in contact with SSgt Penniston and we took a walk through where we saw the lights. Nothing. Finally, we made it back to our vehicle, after making contact with the PC's and informing them of what we saw. After that we met MSgt Chandler and we went in service again after termination of the sighting.

On some of the statements there are handwritten notes, initialed "H." The *H* clearly stands for "Halt," who seems to have written them some years later, in relation to the possibility of the witnesses going public with their experiences. These handwritten annotations give an intriguing insight into the character of the key players and into Halt's mind-set. The remarks are as follows:

Buran: Fred Buran is a good + reliable person. He might talk if his name were protected.

Chandler: No annotation.

Penniston: Sgt Penniston has a lot to contribute. He promised me a plaster cast + photos but never delivered. I think he's holding out to "sell" a story. He is, however, a very competent individual and can be trusted. I'm convinced his story is as he says. He was so shook [up] he had to have a week to recover.
[Penniston loaned Halt one of the plaster casts in 2003 for analysis.]

Burroughs: Burroughs is a straightforward + honest cop. He does have the ability to take an incident + turn it into a disaster (he comes on too strong). There's no doubt in my mind his statement is accurate. He really became obsessed with this. Now he's worried that this might affect his career.

Cabansag: I'm convinced this is a "cleaned up" version of what happened. I talked with Amn Cabansag + can say he was shook up to the point he didn't want to talk. From talking with Chuck de Caro (CNN) I can say he still worries today. He might talk if approached right.

The reference to CNN is another clear implication that these notes were annotated on to the statements some years after the incident, in relation to the question of if/how to go public about the incident. But of particular relevance to the issue at hand is Halt's reference to a "cleaned up" version of what happened.

We have already reviewed Charles Halt's various first-person statements on the events, over the years, including one delivered on November 12, 2007, at a press conference held at the National Press Club in Washington, D.C. Jim Penniston was also present at this event, and because of the importance of quoting primary source material where available it's worth reproducing his position statement here, so we can compare and contrast it with the original witness statement quoted earlier:

> My name is James Penniston, United States Air Force Retired.
>
> In 1980, I was assigned to the largest Tactical Fighter Wing in the Air Force, RAF Woodbridge in England. I was the senior security officer in charge of base security.
>
> At that time I held a top-secret US and NATO security clearance and was responsible for the protection of war-making resources for that base.
>
> Shortly after midnight on the twenty-sixth of December 1980, Staff Sergeant Steffens briefed me that some lights were seen in Rendlesham Forest, just outside the base. He informed me that whatever it was didn't crash . . . it landed. I discounted what he said and reported to the control center back at the base that we had a possible downed aircraft. I then ordered Airman Cabansag, AIC Burroughs to respond with me.
>
> When we arrived near the suspected crash site it quickly became apparent that we were not dealing with a plane crash or anything else we'd ever responded to. There was a bright light emanating from an object on the forest floor. As we approached it on foot, a silhouetted triangular craft about nine feet long by six-point-five feet high came into view. The craft was fully intact sitting in a small clearing inside the woods.
>
> As the three of us got closer to the craft we started experiencing problems with our radios. I then asked Cabansag to relay radio transmissions back to the control center. Burroughs and I proceeded towards the craft.
>
> When we came up on the triangular shaped craft there were blue and yellow lights swirling around the exterior as though part of the surface and the air around us was electrically charged. We could feel it on our clothes, skin, and hair. Nothing in my training prepared me for what we were witnessing.
>
> After ten minutes without any apparent aggression, I determined the craft was non-hostile to my team or to the base. Following security protocol, we completed a thorough on-site investigation, including a

full physical examination of the craft. This included photographs, notebook entries, and radio relays through Airman Cabansag to the control center as required. On one side of the craft were symbols that measured about 3 inches high and two and a half feet across.

These symbols were pictorial in design; the largest symbol was a triangle, which was centered in the middle of the others. These symbols were etched into the surface of the craft, which was warm to the touch and felt like metal.

The feeling I had during this encounter was no type of aircraft that I've ever seen before.

After roughly forty-five minutes the light from the craft began to intensify. Burroughs and I then took a defensive position away from the craft as it lifted off the ground without any noise or air disturbance. It maneuvered through the trees and shot off at an unbelievable rate of speed. It was gone in the blink of an eye.

In my logbook (that I have right here) I wrote: "Speed—impossible." Over eighty Air Force personnel, all trained observers assigned to the Eighty-First Security Police Squadron, witnessed the takeoff.

The information acquired during the investigation was reported through military channels. The team and witnesses were told to treat the investigation as "top secret" and no further discussion was allowed.

The photos we retrieved from the base lab (two rolls of thirty-five-millimeter) were apparently overexposed.

Again, this more fulsome statement, delivered after Penniston retired from the USAF, bears out the point that the original statements were "watered-down" or "cleaned-up" versions of the story. There was just enough basic information in them that they could be said to be essentially true, but with much of the more exotic material not covered we are perhaps, in the territory of a "material omission."

As we look at the five contemporaneous witness statements as a whole, it seems clear that the process here was haphazard to say the least. The statements from Buran and Chandler were typed on an official Air Force Statement of Witness form, signed, and dated, but Cabansag's wasn't on the Air Force form and wasn't dated. Neither was Penniston's (which was also unsigned), while the statement from Burroughs was handwritten, signed, but not dated. In other words, the statements are in a sort of inverse order, where the more closely involved with the sighting the witness was, the less in line with any due process the statement was.

Furthermore, no statements were taken (or, if they were, they have yet to come to light) by some of the other personnel involved in the first UFO sighting, such as Staff Sergeant Bud Steffens, Sergeant "Crash" McCabe, and Sergeant John Coffey.

Moreover, there are no statements from any of those involved in the second night's sighting, such as Lieutenant Colonel Charles Halt, Lieutenant Bruce Englund, Master Sergeant Bobby Ball, and Sergeant Monroe Nevels.

Similarly, there are no statements from some of those more peripherally involved but who would have been able to add important information, such as Captain Mike Verrano and Master Sergeant Ray Gulyas.

This was either an extraordinarily inept investigation or one where some evidence is missing. So what do the personnel who made (or supposedly made) these statements have to say about all this? Buran has said that his statement is accurate and Chandler has not gone on the record about this. Neither men, of course, saw anything themselves. But when it comes to the three men who got closest to the UFO, the story is very different and there are discrepancies.

For example, there was an exchange of remarks directly after the incident, as Penniston and the others prepared to turn in their M-16s. Penniston said to Chandler, "You wouldn't believe what happened out there." Chandler responded in a very sympathetic way, "Yes, I know; I was at the East Gate." This exchange, just before Penniston and Burroughs reported to the shift commander's office, is very different from what Chandler says in his official statement, in summary: "At no time did I observe anything from the time I arrived at RAF Woodbridge."

Cabansag states that he couldn't type and says that someone handed him a typed statement to sign. He says that he was only newly qualified and was extremely nervous. Cabansag was clearly intimidated by all of this and signed, under extreme duress as he puts it, not even having read the text.

In December 2012, to help clarify the position concerning the various post-incident debriefings and the witness statements, Burroughs and Penniston jointly wrote up the following account of what happened:

I was methodically and consistently interviewed and interrogated by my chain of command and other agencies. Every time, I was promised that this was the last interview and it would be absorbed into the classified annals of data and I would need not tell or talk about it no more. This was not the case. I went through at least fourteen debriefings and two by non–air force personnel. I gave all information from memory and at no time was the notebook ever brought up. The debriefs were all for the last time, I was promised. Tell all, and tell it correctly, and it would be the last of questions on Rendlesham. For these were to continue, no matter what I had said. I do believe the command element, were more for obtaining knowledge. But the external interrogation, were for much more, I am afraid.

The timeline for Rendlesham and information flow: incidents occurred Dec 26–28, 1980. On the morning of the 29th of December, AFOSI building, meeting with two American Agents, more likely Defense Intelligence Agency (DIA) and/or National Security Agency (NSA), Penniston writes a four page written statement to the agents. He dates it and signs the document. Agents then give Penniston a typed statement, which is generic and is limited on details. For example, observation of a metallic craft and not getting within 50 yards of it. Penniston is instructed by the agents that an official investigation is underway, and he is to tell all who ask the cover story that was provided to him. He reads it several times and then agrees to do so. I go up to the Wing and Base Command offices; as I walk in Colonel Conrad motions for me to come in to his office. He shuts the door; here I am with my Base Commander; he asked me some questions, but not as a commander, more like a fellow airman and he also gave me encouragement to understand we had his support. I then go over to the Deputy Base Commander's office down the hall. Burroughs is waiting outside Halt's office and meets him and we both ask Halt's secretary to tell him we are there. Then we are debriefed at the Deputy Base Commander's office, Colonel Halt. Statements written and then drawings made. Penniston, Burroughs and Cabansag are taken into Wing Commander's Office with Base and Deputy Base Commanders present. The NSA account is briefed to the officers. The Wing Commander thanks the Security Policemen for the report and asks no questions.

All direct witnesses are briefed to treat all discussion about Rendle-sham as Top Secret. Ending all conversation on base by those directly involved. We are assured that this incident will never see the light of day. For it is classified and names have been sanitized from reports. The most unusual thing was after I give Colonel Williams the sani-tized version that was provided down at the AFOSI building. Some-thing so unusual struck me as Colonel Williams patiently listened to our account of the happenings of the first night. The account never stirred one question at all, nor comment. He merely thanks us for do-ing our job, and he appreciated the report. I have always thought, would there not be just one question? It was if he knew about the phe-nomenon that I had just briefed him on.

Penniston offered this even more damning assessment of the duplicity in September 2013:

Any time a statement of witness is done, it is usually done on a form, or if not available, on lined paper. All witness statements should be dated, signed, and witnessed. If they are not, it should be questioned as to whether they were actually made by the witness. I personally believe the only statement that was actually authentic and not tam-pered with is the one from John Burroughs. I was never asked by the Security Police Squadron to do a statement. This is because they knew it had already been accomplished at the base AFOSI building. Bur-roughs' statement was asked for by Colonel Halt and not by the Secu-rity Police Squadron. I believe that the other statements were either the product of coaching, or written by others, which they had the wit-nesses sign. Case in point, Ed Cabansag's statement; it was prepared and he was told to sign it. So was there containment? Absolutely; so they could develop the cover story of aliens and of fixing an alien spacecraft using A-10 avionics—almost laughable that anyone would buy that rubbish. I guess the bottom line is that "just another UFO story" would develop from the scuttlebutt and from the story put out in the local pubs and on base. Anyone who was a witness or in the know about the incidents was silenced and the people who were talk-ing were people who unknowingly to themselves perpetuated the cover story, some of them by proclaiming that they were witnesses to the events.

Burroughs, too, reiterates the key point that the so-called witness statements were essentially works of fiction, downplaying the key events: "That's the statement they asked me to write. All they wanted was a brief overview on what happened the first night, and that's what I did."

Perhaps the most damning remarks about the way in which the witnesses were treated in the aftermath of the Rendlesham Forest incident come from Charles Halt. Writing in a chapter of Leslie Kean's book *UFOs: Generals, Pilots, and Government Officials Go on the Record*, Halt made the following extraordinary statement: "OSI [Office of Special Investigations] operatives harshly interrogated five young airmen, some of them in shock at the time, who were key witnesses." Halt went on to write: "Drugs such as sodium pentothal, often called a truth serum when used with some form of brainwashing or hypnosis, were administered during these interrogations, and the whole thing has had damaging, and lasting, effects on the men involved." Although the "five young airmen" are not named, it seems clear that this is a reference to the five witnesses from whom written statements were taken.

While some might regard this as "too little, too late," and would doubtless have expected senior officers to have stepped in to protect their men at the time, Penniston is pragmatic about this and grateful that Halt spoke out at all. Penniston offers this view, ending with a speculative bombshell: "I feel there was a whole array of things that Colonel Halt did not fully know at the time. However, through the course of time he did become aware and finally acknowledge the use of drugs and interrogation of the first responders. Maybe, just maybe, Halt was interrogated also. I often wondered about that, especially given his gaps in memory too." Whatever one believes about UFOs, for a senior military officer to make the allegation that drugs and hypnosis were used on the men under his command in the immediate aftermath of a UFO incident is little short of sensational.

8. THE BRITS ARE COMING

At some point on Sunday, December 28, 1980, Lieutenant Colonel Charles Halt (who had grabbed a few hours of sleep after his investigation and sighting on the night of December 27/28) had a conversation about the incident with the Wing Commander, Colonel Gordon Williams, who was the commanding officer of the twin bases. Williams asked Halt to brief him, which he did. He also played Williams the tape that he'd recorded. Williams asked to borrow it, with the intention of playing it to the senior USAF officer in the United Kingdom, General Robert W. Bazley. Although this was phrased as a request, it was clearly not one that Halt could decline and he duly handed over the tape. As he recalls, he was worried that his next promotion and perhaps even his security clearance were in jeopardy (fears that proved groundless) and he spent the next couple of days worrying what the fallout would be.

On Tuesday, December 30, Williams went to RAF Mildenhall, where Bazley was based. The USAF presence in the United Kingdom is known as 3rd Air Force and Bazley—as commander of 3rd Air Force—held a regular staff meeting with the commanding officers of the various USAF bases in the United Kingdom. These meetings were a two-way street. They provided an opportunity for base commanders to brief and discuss their big current issues in one another's presence and with Bazley and other key HQ staff from 3rd Air Force. Conversely, it was an opportunity for Bazley and his top team to brief his various commanders about wider USAF and DoD issues that might impact on their work.

At some point during this meeting on December 30 Williams briefed Bazley, Bazley's key staff, and the other COs about the UFO sighting. Halt's tape was played. According to Halt, who wasn't present but whom Williams briefed later, there was a stunned silence. Bazley raised his eyebrows and asked his assembled staff what the hell they should do now. An unnamed officer apparently made a remark along the lines of, "Wow, this thing's bigger than the . . . [Roswell?] affair"—though, frustratingly, the penultimate word is not recalled with 100 percent accuracy.

At that point, Bazley asked for clarification on where, precisely, the initial UFO incident had occurred. When Williams confirmed that it was in the forest, i.e., "outside the wire," Bazley looked pleased and uttered words to the effect that "it's a Brit affair. It happened off the installation. Let them handle it." Penniston is clear in his view that this was a cunning sleight of hand: "This was the Air Force's way to fulfill the containment of the incident. A clever smoke screen that fitted nicely within the parameters of operating procedures. The end result would make one think the other is investigating the incident." Burroughs agrees but actually welcomed the move: "It was a smart move, since all of what happened was off-base. Yes, it was a smoke screen, but it took a lot of heat off of us early on."

When Williams returned from the meeting he told Halt what Bazley had said and suggested that he liaise with the RAF Commander, Squadron Leader Donald Moreland, who served as the liaison officer between the twin bases and the UK authorities. As it happened, Moreland was on Christmas leave and Halt was not able to discuss the issue with him until around Monday, January 12. Halt recalls that Moreland told him to write a memo to the UK MoD but to "sanitize" it. In later years, Halt described this "sanitization" in the following terms: "The memo was not meant for public consumption. It was meant as a tickler, if you will, to the Ministry of Defence, to get them involved to do a proper investigation."

So now we come to one of the most critical documents in relation to this case, the so-called Halt memo. Given the events that have been previously described, it's as interesting for what it doesn't say as what it does, another "material omission," perhaps. It's also important because it was the first formal notification to the UK MoD and later, the first document relating to the case to be made public.

The Halt memo was dated January 13, 1981, and in line with the "sanitization"/"tickler" strategy went out under the understated title "Unexplained Lights." The memo reads as follows:

1. Early in the morning of 27 Dec 80 (approximately 0300L) two USAF security police patrolmen saw unusual lights outside the back gate at RAF Woodbridge. Thinking an aircraft might have crashed or been forced down, they called for permission to go outside the gate to investigate. The on-duty flight chief responded and allowed three patrolmen to proceed on foot. The individuals reported seeing a strange glowing object in the forest. The object was described as being metallic in appearance and triangular in shape, approximately two to three meters across the base and approximately two meters high. It illuminated the entire forest with a white light. The object itself had a pulsing red light on top and a bank(s) of blue lights underneath. The object was hovering or on legs. As the patrolmen approached the object, it maneuvered through the trees and disappeared. At this time the animals on a nearby farm went into a frenzy. The object was briefly sighted approximately an hour later near the back gate.

2. The next day, three depressions 1 ½" deep and 7" in diameter were found where the object had been sighted on the ground. The following night (29 Dec 80) the area was checked for radiation. Beta/gamma readings of 0.1 milliroentgens were recorded with peak readings in the three depressions and near the center of the triangle formed by the three depressions. A nearby tree had moderate (.05–.07) readings on the side of the tree toward the depressions.

3. Later in the night a red sun-like object was seen through the trees. It moved about and pulsed. At one point it appeared to throw off glowing particles and then broke into five separate white objects and disappeared. Immediately thereafter, three star-like objects were noticed in the sky, two objects to the north and one to the south, all of which were

about 10° off of the horizon. The objects moved rapidly in sharp angu-
lar movements and displayed red, green and blue lights. The objects to
the north appeared to be elliptical through 8–12 power lens. They then
turned to full circles. The objects in the north remained in the sky for
an hour or more. The object to the south was visible for two to three
hours and beamed down a stream of light from time to time. Numer-
ous individuals, including the undersigned, witnessed the activities in
paragraphs 2 and 3.

Despite the massive downplaying of the incident, it was still pretty
sensational stuff, with its reference to an "object" described as being "metallic in appearance and triangular in shape," the account of the radiation
readings taken at the landing site, and the admission at the end that Halt
himself was one of the witnesses—though he could hardly have downplayed his involvement any further without editing himself out of the
story altogether!

Should Halt have said more? Penniston thinks it struck the right balance: "Colonel Halt was respecting our privacy, as he had guaranteed. Also,
he believed it would never see the light of day. I think it said enough."

Moreland forwarded the memo to an MoD division called Defence Secretariat 8 (DS8) two days later. His brief January 15, 1981, cover note went
out under the title "Unidentified Flying Objects (UFOs)" and read as follows: "I attach a copy of a report I have received from the Deputy Base
Commander at RAF Bentwaters concerning some mysterious sightings in
the Rendlesham Forest near RAF Woodbridge. The report is forwarded for
your information and action as considered necessary."

The UK MoD is broadly equivalent to the US DoD. It has a dual role as
a policy-making Department of State and as the United Kingdom's highest-
level military headquarters. This can lead to some confusion but is important in understanding where the MoD sits within the British government.

Broadly speaking, the MoD's involvement with the UFO issue mirrored
that of the old USAF program Project Blue Book. The remit was to look at
UFO sightings reported to the MoD, investigate them, and determine
whether or not there was evidence of any potential threat to the United
Kingdom or anything of more general defense interest. These two possible
outcomes were sometimes lumped together using the phrase "defence

significance." The methodology of the MoD's UFO investigations and indeed the conclusions also mirrored those of Project Blue Book, and it's clear that in setting up the United Kingdom's program some aspects of strategy, structure, and process (right down to the design of the forms used to record sighting reports) were based on the Blue Book model. Bizarrely, unlike its US counterpart, the UK program had no formal name, though the British media often uses the phrase "MoD's UFO project" as a clear and concise descriptor. Over the years this work has been undertaken by a wide range of different MoD sections, with titles including S4, S6, Defence Secretariat 8 (DS8), Secretariat (Air Staff) (Sec(AS)), and Directorate Air Staff (DAS), to name but a few. Clearly, this "alphabet soup" is another reason why the media's "MoD's UFO project" tag is appropriate.

Given that the initial UFO sighting happened in the early hours of December 26, the fact that the date of Halt's memo to the MoD is January 13 is astonishing. Furthermore, given that the covering note from Donald Moreland was dated January 15, it's quite possible that Halt's memo didn't reach the MoD's UFO project until the week commencing January 19. While the reasons for this have been explained, any police officer will tell you that the first twenty-four hours (maybe forty-eight, in some cases) are absolutely critical in terms of an investigation and that if significant progress isn't made within this period the chances of success diminish rapidly. The huge delay in informing the MoD fatally undermined the United Kingdom's ability to conduct a meaningful investigation and meant that to all intents and purposes only the United States stood a chance of successfully resolving the mystery in the immediate aftermath of the events.

There was to be one other consequence of the delay, in that it gave the MoD a useful "get out" in later years, when responding to media and public inquiries about the events. By pointing out that Halt had waited nearly three weeks before reporting matters to the United Kingdom, the MoD was able to say that the US authorities clearly had matters well in hand and/or were relatively unconcerned about the events. This allowed the MoD to imply that the events were of little consequence, an implication that was entirely consistent with the long-standing MoD policy to downplay both the UFO phenomenon itself and the level of official interest and resources expended on the subject.

As if this delay weren't bad enough, the memo contained a critical factual error that was to have far-reaching consequences: the memo described the first sighting as having taken place "early in the morning of 27 Dec 80" when, in fact, the correct date was December 26. Consequently, when the memo set out the other key events, prefaced with the phrases "the next day," "the following night," and "later in the night," everything was wrong, incorrectly reported as having taken place twenty-four hours later than was, in fact, the case. One obvious problem that this caused was that when, later on, the MoD came to check whether anything unusual had been tracked on radar the wrong dates were checked. Penniston wonders if this was deliberate: "Yet another enigma with the Rendlesham Forest incident. Why the delay and the inaccuracy with the dates? I believe it was to help with the containment. By incorrect times and dates, it is hard to investigate at a later date through FOIA."

Burroughs, too, speculates that this was part of a deliberate strategy: "The fact he got the dates wrong in the memo was not by accident." Burroughs, however, paints Halt as a victim, too, and not as the perpetrator: "I believe it was written in case the incident got too hot; that Halt—because he was out there—would be the fall guy as far as taking the heat for it and having to answer all the questions. They would say it was just strange lights seen in the dark and show we were not covering anything up. The dates were wrong to keep people from getting any information under FOIA."

It's likely that the desk officer on the MoD's UFO project didn't receive Halt's memo and Moreland's covering note until the week commencing January 19—fully twenty-four days after the first events. Notwithstanding this delay, the MoD's UFO project had a huge advantage in relation to those USAF personnel caught up in these events, given that the USAF's Project Blue Book had been terminated in 1969 and was probably unfamiliar to most—if not all—of the personnel involved in the Rendlesham Forest incident. Conversely, when Halt's report did finally get to the MoD, it was sent to a section that had considerable experience in dealing with UFO sightings, access to a wide range of investigative resources in terms of both equipment and personnel, and an archive of hundreds of files on the subject. This should have ensured they were well placed to get things rapidly back on-track, despite the fact that they came late to the game.

Ironically, this actually worked against those concerned, probably because just about all previous sightings had involved aerial encounters rather than landings. Consequently, a tried and tested process for investigating such cases was applied to a totally different situation, where an altogether different approach should have been taken. That said, the desk officer got two things right: First, a key priority was to check whether radar evidence might corroborate what had been reported. Second, Halt's memo contained data on the radiation readings taken at the landing site and the desk officer knew that the MoD had specialist staffs that could make an informed assessment of this.

We've previously made it clear that both the United States and the United Kingdom had jurisdiction. A separate question, however, is who had primacy. This is important because the answer to this question should have determined whether the United States or the United Kingdom took the lead in investigating and handling this incident.

However, for governments the UFO issue is—at best—something of a hot potato. Charles Halt has described it as a "tar baby." Consequently, the confusion over primacy has been quite useful to both the United States and the United Kingdom, with each implying (and sometimes openly stating) that primacy lay with the other! The MoD has been able to say that the USAF's delay in reporting showed that the US authorities had matters well in hand. Conversely, when briefed about the incident by Gordon Williams, Robert W. Bazley could barely disguise his relief when told that the key events had taken place off the bases: "It's a Brit affair."

One consequence of the confusion over jurisdiction and primacy was poor information sharing. The USAF didn't pass the witness statements that Halt took from five of the key witnesses to the MoD. Critically, this included the sketches of the craft and the symbols made by Jim Penniston, which should have set alarm bells ringing with all concerned. This failure, coupled with the delay in reporting the incident to the MoD, was clearly a factor in the MoD's failure to conduct any follow-up interview with Halt or the other witnesses—itself another key failure. Conversely, the MoD failed to pass to the USAF the Defence Intelligence Staff (DIS) assessment of the radiation levels at the landing site. The consequence was that each party had information that would have been useful to the other and would clearly have demonstrated that the incidents were far more significant than was realized at the time.

The most controversial instance of this lack of information sharing relates to one of the most extraordinary parts of this whole affair—the involvement of General Charles A. Gabriel, then CINCUSAFE (Commander in Chief US Air Forces in Europe). In a document dated February 16, 1981, Squadron Leader Badcock, an RAF officer specializing in air defense issues, wrote to the MoD's UFO project about the case. After addressing the issue of whether or not the UFO was tracked on radar, he finishes by writing the following extraordinary sentences: "I asked if the incident had been reported on the USAF net and I was advised that tape recorders [sic] of the evidence had been handed to Gen Gabriel who happened to be visiting the station. Perhaps it would be reasonable to ask if we could have tape recordings as well." This is worthy of further comment. The first point to make is that "tape recorders" is almost certainly a reference to the audiotape recorded by Lieutenant Colonel Charles Halt on the second night of activity. Clearly, however, it would be odd if the general took possession of this in isolation. It's far more likely that this formed part of a more extensive briefing package including, for example, the soil and sap samples taken from the landing site, which have disappeared without a trace. But this raises wider questions about the general's visit. At first glance, the phrase "who happened to be visiting the station" implies that this was a pre-planned "meet and greet" visit and that any briefing on the UFO sightings took place in the margins of this routine visit. But General Gordon Williams is insistent that Gabriel's "meet and greet" visit took place before Christmas—and thus before the UFO sightings. It seems then that the visit referred to in the MoD document was a subsequent visit, details of which had not been briefed to Williams.

The possibility that General Gabriel made a second visit to the twin bases so soon after his routine pre-Christmas visit is problematic, as is the suggestion that Williams, the commanding officer, wasn't briefed on this. But if Williams is correct about the timing of the "meet and greet" visit, the MoD document does indeed suggest that a separate, secret visit took place, in direct response to the Rendlesham Forest incident.

Again, while rumors of unscheduled visits to the base in the aftermath of the events are widespread, it's Halt who is clearest on the issue. He has stated that a C-141 transport aircraft arrived but that there was great secrecy about this and he was not able to ascertain who was onboard and what their mission was. Apparently a group of "special individuals" disembarked, passed through the East Gate, and headed out into the forest. Was this the flight on which General Gabriel arrived? And if so, why were Williams and Halt apparently not involved or even briefed? Penniston has his view on this: "My thoughts are that what happened at the twin bases was so important and retrieval of documents and other evidence was so paramount, it was only entrusted to a four star general, the Commander-in-Chief of USAF Europe. Someone thought it was necessary to have General Gabriel handle it and that no delegation would be allowed to anyone else. This order came from the top, I suspect." Burroughs agrees wholeheartedly on this point: "That shows just how serious the incident really was, for him to get involved; and at what level it finally ended up at."

A key point in every MoD UFO investigation was an examination of military radar data. Indeed, UFO reports were sent automatically to the Directorate of Air Defence, a division in the MoD that had responsibility for Air Defence Ground Environment (ADGE) issues. This was done so that UFO sightings could be cross-checked with radar data to see if anything had been tracked that might corroborate a visual sighting. A negative result didn't mean sightings were dismissed out of hand—stealth technology illustrates that solid objects can have a very low (ideally zero) radar signature—but radar was a key factor in assessing UFO sightings as "explained" (i.e., misidentifications of ordinary objects or phenomena), "unexplained," or cases where there was "insufficient information" to make a meaningful assessment.

On January 26, Squadron Leader Badcock wrote to two radar units—RAF Neatishead and Eastern Radar, based at RAF Watton—asking them to check whether any unusual radar readings had been recorded on the evening of December 29, the date Halt had given for the second night of activity. But as we've seen, this date was wrong. To make matters worse, for some reason Badcock omitted to ask that the radar data for the first night of activity (December 26, wrongly described by Halt as December

27) be checked when, arguably, this was the more significant of the two nights, given that the UFO actually landed.

Squadron Leader Sharpe from RAF Neatishead replied on February 5, stating that the radar camera recorder (two radar screens displayed identical data—one was monitored by a radar operator and one was filmed, so that a record would exist of the data) had been switched off earlier in the day, once military flying activity had finished, as was normal practice. Sharpe went on to say: "An examination of executive logs revealed no entry in respect of unusual radar returns or other unusual occurrences."

Eastern Radar at RAF Watton replied on February 26. Squadron Leader Coumbe stated that the film of their radar camera recorder for the evening of December 29 was faulty. Helpfully, they then checked the film in respect of December 28 and December 30, but both these films were faulty, too. Intriguingly, Coumbe went on to say: "On the night of the reported sighting our controller on duty was requested to view the radar; nothing was observed. The facts are recorded in our log book of that night." Coumbe does not give the date, so it's not clear whether this is a reference to the first or the second night of activity.

As can be imagined, the fact that the Neatishead radar camera recorder was switched off, the Watton radar camera recorder films were faulty, and, in any case, the wrong days were checked has led to some conspiracy theories. At the very least, it was a sorry state of affairs, perhaps best summed up in an undated internal note sent to the UFO project around four years later, by an officer in the ADGE division, Wing Commander Keith, who wrote: "Regrettably, the tasking letter from MoD referred to an incident on 29 Dec 80 therefore the replies from Neatishead and Eastern radar are probably worthless. Unit radar recordings are not held for 4 years consequently we are back where we started!" On a point of clarification, standard procedure at the time was to keep paper records of radar data for three years before destroying them. Video recordings of radar data were retained for thirty days, prior to re-use of the tapes.

The point about the duty controller at Eastern Radar being asked to look at the radar on the night of the sighting is worthy of further comment. Despite Squadron Leader Coumbe's assurance, the radar operator he refers to—Nigel Kerr—tells a different story. Kerr does not recall the exact date but says that sometime around the Christmas holidays a call came in from Bentwaters about "a flashing light in the sky." He checked the radar and saw something on the approach line, which he first thought was a helicopter. It remained stationary for three or four sweeps across the radar screens before it dissipated. It's not clear whether or not this ties in with an entry in the RAF Watton log timed at 0325 on December 28, which states: "Bentwaters Command Post contacted Eastern Radar and requested information of aircraft in the area—UA37 traffic southbound FL370—UFO sightings at Bentwaters. They are taking reporting action." Is this the call Kerr recalls, or was this a second call, placed when Kerr was off-duty? And was this call the one to which Squadron Leader Coumbe referred in his February 26 letter when he wrote: "On the night of the reported sighting our controller on duty was requested to view the radar . . ."? Frustratingly, it's not clear.

It has been variously suggested that in the immediate aftermath of the events unidentified US officers visited RAF Watton to examine and/or to remove radar recordings. There are several such accounts, but it is difficult to nail down the facts. Two particularly intriguing pieces of testimony relate to two USAF air traffic controllers—Ivan "Ike" R. Barker and James H. Carey—who apparently tracked a UFO on radar at Bentwaters at some point between Christmas and New Year. But again, the facts are difficult to pin down.

Aside from radar data, the second substantive issue that the MoD's UFO project looked into was radiation. Lieutenant Colonel Halt's memo to the MoD described how radiation readings were taken at the landing site, at his behest, by Monroe Nevels from the Disaster Preparedness Office.

The readings in Halt's memo were passed to officers in the DIS who were asked to comment. The DIS was one of a number of areas that the MoD's UFO project could call upon for specialist advice and assistance. The usual port of call was DI55, which reported to DGSTI—Director General Scientific and Technical Intelligence. On this occasion, because of the radiation issue, DI55 brought in DI52, another part of DGSTI's empire.

At the time of the Rendlesham Forest incident and indeed for many years thereafter, the role of the DIS in relation to UFO research and investigations had not been publicly acknowledged. The UFO community, however, suspected their involvement. Sometime in the eighties an apparently innocuous photocopy of a relatively unexciting UFO report was sent to a member of the public by a helpful desk officer on the UFO project. But somebody had forgotten to black out the internal MoD distribution list (as was the usual practice), where, among other divisions, DI55 was listed. It didn't take a genius to figure out what "DI" stood for! Notwithstanding this slip, the involvement of the DIS in MoD UFO research and investigations was not formally acknowledged until comparatively recently.

On January 28, Squadron Leader Badcock, acting on a request from DS8, wrote to DI55 and among other things asked: "We would particularly like to know whether the readings of radioactivity are unusual or whether they are in the normal background range to be expected."

DI55 passed this to DI52. On February 23 they replied (it's not clear why this took so long), stating: "Background radioactivity varies considerably due to a number of factors. The value of 0.1 Milliroentgens (mr), I assume that this is per hour, seems significantly higher than the average background of 0.015 mr."

In the course of a 1994 cold-case review of the Rendlesham Forest incident the MoD rechecked the issue with the Defence Radiological Protection Service. It was confirmed that the radiation levels reported were somewhere between seven and ten times higher than expected background levels. It was stressed that while statistically significant, such levels of radiation would not be harmful to those concerned. Notwithstanding this, given that we don't know anything about the source of this radiation or the type, the issue is of particular concern to the witnesses, particularly John Burroughs and Jim Penniston, who believe their current adverse health conditions are directly attributable to the events of December 1980. Clearly, the question of health problems caused by Burroughs and Penniston being irradiated is an issue on which the chain of command is potentially extremely vulnerable. But would this failure rest with the MoD or the USAF? While there's no documentary evidence that the MoD briefed the USAF on a potential radiation hazard to their men (which would have been a catastrophic failure in and of itself), Burroughs cannot conceive that this information would not have been briefed, at some level: "I'm not sure they [the MoD] didn't brief the USAF on this." Either way, he's clear on the failure to alert those most directly involved: "The fact that they [the MoD and/or the USAF] held back information that affected the people who responded out there is criminal."

Some skeptics in the UFO community have suggested that the radiation levels might not be as significant as the MoD suspected, arguing that the Geiger counter used was not appropriate for the task and even speculating that the dial might have been misread. I'm wary when ufologists start trying to second-guess the measurements taken by the trained military personnel who were actually there or questioning the contemporaneous scientific assessment. Nevels used the equipment available to him (there being no such thing as a UFO radiation detector!) and the DIS assessment used the readings reported to the MoD. We can only use the data we have, not the data we'd like to have or think we should have had. That's the way science works. In any case, such speculation misses the key point; the radiation levels peaked in the three indentations found where the craft was said to have landed. It's like using a metal detector and hearing a bleep; in a sense, it doesn't matter what make or model of metal detector it is or whether its dial reads 7 out of 10 or 8 out of 10; the key point is that it bleeped—that tells you there's something there!

Despite the USAF's delay in telling the MoD about the sightings, despite the fact that the incorrect dates were notified, and despite the problems with the radar camera recorders/films, matters were still retrievable. As a result of the MoD's initial investigation, three follow-up actions were now appropriate—one was a matter of common sense, while the other two had been the subject of specific suggestions and offers.

The commonsense action was for someone at the MoD to pick up the phone and speak to Donald Moreland, asking him to facilitate a meeting with Halt and—perhaps—some of the key witnesses such as Burroughs and Penniston. This is what Halt had expected and he was surprised when it didn't happen.

The two areas where specific suggestions/offers had been made related to Halt's audiotape and to the radiation readings.

First, Squadron Leader Badcock's February 16 letter to DS8 had concluded by saying: "Perhaps it would be reasonable to ask if we could have [these] tape recordings as well." Readers may sense a certain frustration in Babcock's words. Given the unwritten "Don't criticize the Americans" rule that the MoD, and arguably the UK government as a whole, operated on (the United Kingdom is sometimes dubbed America's poodle) and given the understated and diplomatic language that UK government officials often use, it should be clear that the casual sentence quoted here masks the absolute fury that the MoD felt at the USAF for having removed evidence from the country without briefing them.

Second, the DI52 letter containing the scientific assessment of the radiation levels included the following offer: "If you wish to pursue this further I could make enquiries as to natural background levels in the area."

Why were these three actions not taken? In relation to the failure to follow-up with Halt, it may be that as the MoD subsequently claimed, his initial delay in reporting sent (wrongly) the message that the US authorities had matters in hand and/or were not overly concerned by the events.

So far as the tape recordings are concerned, the issue was raised again in 1983 and the ADGE division explained to DS8 why no action had been taken, stating: "At Reference you ask if the suggestion that the USAF be asked for the tape recordings was followed up by this Deputy Directorate. It was considered that the tapes would reveal no better report than that already reported, and no further request was made." While it may be the case that there was some confusion over whether DS8 or the ADGE division should have been taking this action, this is a sorry state of affairs, to say the very least.

As for the failure to take up DI52's offer concerning the radiation readings, this is equally baffling. One possibility has to do with the MoD's consistent policy of downplaying their involvement with UFO investigations and constantly pushing the "no defence significance" line when the subject was raised in Parliament or by the media and the public. This led to a natural defensiveness on the part of some officials, worried that any proactivity in relation to UFO research might somehow compromise the "no defence significance" public line. But this is weak. The United Kingdom didn't get a Freedom of Information Act until many years later and officials writing in 1980 knew—or should have known—that under the law as it stood at the time, documents they wrote wouldn't be made available to the public for at least thirty years.

Finally, it's also surprising that no internal MoD meeting seems to have taken place. Various parties were writing to one another, but there's no evidence that staff in DS8 ever called a meeting with the various DIS and ADGE staffs who had been looking into the events. If anything, the MoD has a "culture of meetings," many of which are superfluous or of limited value. Here was an instance where a meeting could genuinely have added value and helped those concerned decide how best to proceed. So far as we're aware, no such meeting ever took place.

Jim Penniston is scathing about the MoD's handling of all this:

I do believe the MoD thought they were the office of primary responsibility. They eventually wished it was an American problem, because they had a lot to answer for. So silence was their defense, say nothing and hopefully it would go away. What made it even more imperative not to talk to the US was the discovery of the high radiation level, not to mention the radar tapes, which had gone missing. The MoD was confronted with the stark reality of an unknown craft which was recorded on radar and now the high levels of radiation on the ground showed it landed too. Now there was no room for debate within the Ministry. The reality is that over the course of three nights in December 1980, while the population of England slept, it has become apparent the MoD was sleeping as well. The airspace was open over the UK for the three nights in question. So how do they explain the incompetence of the MoD? A craft of unknown origin entered UK airspace twice and left in the same manner over the course of the three day period. They were detected by radar and now there was clear evidence a touchdown and takeoff occurred, while the MoD slept, leaving the population of England quite possibly unprotected and definitely unaware. They did not even generate a phone call, or generate an aircraft response flight. They were clearly asleep at the wheel. As for us and the other personnel affected by this phenomenon, we were thrown under the bus by the MoD.

Before moving on, we should summarize briefly the areas dealt with earlier. A combination of delay, confusion over jurisdiction/primacy, and poor information sharing led to a fatally flawed investigation—or, rather, investigations. US and UK authorities were both struggling to deal with a situation not catered to by any SOPs or other official guidance. Nobody was clear who was in overall charge and nobody seemed willing to take ownership of the situation; quite the opposite, in fact: the United States was overly keen to portray this as "a Brit affair," while MoD officials took the opposite view.

As for the rest of it, it reads like a comedy of errors: the United States gave the MoD the wrong dates; the MoD forgot to check the first date and find that the radars were switched off during the second. When they did manage to check a date, they found that the film hadn't come out.

When taken collectively, none of this makes happy reading. To be fair, what looks like conspiracy often turns out to be bureaucracy. However, at the very least, the mistakes made represent a colossal missed opportunity. Some would doubtless suggest that the whole catalogue of errors is so unlikely that some other hidden hand must have been involved, conspiring to ensure that the official investigation turned up nothing and that the initial investigators were "set up to fail."

9. SKEPTICAL THEORIES

The time has come to examine the various theories that have been put forward over the years. These theories fall into two categories, which might broadly speaking be labeled "believer theories" and "skeptical theories." Some theories can be dismissed more easily than others, but readers may or may not agree with the various assessments here, so the purpose of these next two chapters is not to attempt to push any one particular theory but simply to expose all the various possibilities that have been discussed and debated over the years—along with a few new ideas.

First, some words of warning. Many of the witnesses have their own views on what they saw and experienced. For example, as we have seen, Charles Halt has clearly stated that he believes the UFO he saw was extraterrestrial in origin. It's tempting for believers to say things like, "Well, he was there," as support for this theory. Fair enough. But this doesn't invalidate other theories and indeed there's a counter-argument that says witnesses, in a sense, are too close to these events to take an objective view; they have too much "emotional investment" and are incapable of taking a step back, looking at the bigger picture, and coming to a more dispassionate assessment. Both arguments have merit, so there's no right or wrong here, but it's worth bearing these points in mind.

Another problem with giving too much weight to the witnesses' theories is that there were two main but entirely separate events and numerous people who were involved in one, the other, or both—to varying degrees. So it's no surprise that different people have different experiences and therefore different theories about what actually took place.

We should be similarly wary of what the UFO community thinks of the Rendlesham Forest incident. Ufologists (as they label themselves) fall into two broad camps: true believers and die-hard debunkers. Just as fascist and communist regimes are actually very similar, despite the apparent distance implied by phrases like "left-wing" and "right-wing," so true believers and die-hard debunkers are actually chiseled from the same block, united by their dogma. Both groups seize on the Rendlesham Forest incident to support their pre-existing beliefs—a classic case of what the scientific community calls confirmation bias. True believers want Rendlesham to give them definitive proof of extraterrestrial visitation, while die-hard debunkers want an explanation that will leave humanity reassuringly alone and unvisited. Any research or investigation the UFO community undertakes tends to be conclusion led. Thus, both the true believers and the die-hard debunkers see what they want to see when looking at these events, as opposed to taking a truly scientific approach and going where the data take them—even if this confounds their personal beliefs and expectations.

With these warnings in mind, let's run through the main skeptical theories.

DRINK AND DRUGS

The most serious allegation about the incident is that alcohol and/or drugs were involved. It's difficult to see how this would translate into the detailed, multiple-witness accounts that we have from two separate nights, involving not just junior ranks but also senior officers, including the Deputy Base Commander. Alcohol simply doesn't generate hallucinations of this kind and it would be straining credulity beyond breaking point to suggest that drugs—even hallucinogenic ones—would result in a shared hallucination involving so many people at so many times.

Few people have seriously suggested this possibility and, ironically, the idea of a hallucination seems inadvertently to have been planted by one of the main proponents for an extraterrestrial explanation, the United

Kingdom's former Chief of the Defence Staff Lord Hill-Norton. In a sound bite on the case that he deployed in a number of similar versions, Lord Hill-Norton often said things like this: "Either large numbers of people were hallucinating, and for an American Air Force nuclear base this is extremely dangerous, or what they say happened did happen." It was certainly not Lord Hill-Norton's view that "large numbers of people were hallucinating," and in fact he deliberately chose an alternative that he regarded as absurd to make the point that in his view the second possibility (i.e., that the events took place as the various witnesses claimed) was correct.

It would be foolish to suggest that excessive drinking and illegal drug taking never occurred at Bentwaters and Woodbridge. The fact that the US military had and still has programs to deal with drug and alcohol abuse shows there's a problem. However, there's no evidence to suggest any of the witnesses whose testimony we've highlighted were drunk or had taken illegal drugs.

It's also worth bearing in mind that most of the witnesses were security police and law enforcement personnel, who might well have to deal with the occasional "drunk and disorderly" but are less likely than most to be abusing drugs and alcohol themselves. Several other factors make the theory unlikely.

First, there's the "buddy system," whereby people often operate in pairs so that they can monitor each other's behavior, with one taking action if the other has a problem. This is both informal and formal—some areas of the twin bases could only be patrolled in pairs, as there are strict rules preventing single individuals from being in proximity to the most sensitive locations, such as the WSA. Second, most of the witnesses were in positions where they would be routinely checked for any indications of drink or drugs before going on duty. "We were checked prior to going on duty for drugs/alcohol, so not possible," as Burroughs put it. Third, though for security reasons we can't go into the details, many of the key witnesses were part of the Nuclear Weapon Personnel Reliability Program (PRP) and were thus vetted, trained, and monitored to an extent that did not apply to other military personnel. Any sign of drug or alcohol abuse (or indeed any other signs of instability) would have been quickly noticed and would have had immediate consequences.

HALLUCINOGENIC MUSHROOMS

To stay for a while with the theory about some sort of hallucination, is it possible that hallucinations were caused by "shrooms" in Rendlesham Forest? Given that the forest is central to the key encounters, could all of the witnesses somehow have been exposed to psychedelic mushrooms? Clearly, mushrooms with psychoactive properties can cause mystical experiences and many people deliberately take them for precisely this reason. Throughout history, people—often shamans—have used certain mushrooms to facilitate a visionary state. People talk about "the shamanic journey" and some people believe that the human mind is capable of accessing some other, hidden realm. Such speculation, which raises wider issues on the nature of consciousness, is outside the scope of this book, but the question of whether mushrooms played a part in the Rendlesham Forest incident is a fair one. The short answer to the question is no. There are certainly plenty of mushrooms and other fungi in Rendlesham Forest, but we're not aware of any "magic mushrooms." And even if there were, it's simply not possible for people to inadvertently ingest the hallucinogen, e.g., via spores blowing on the wind, in a way that would have any discernible effect.

A variation on this theory has been put forward in respect of another famous British UFO sighting. On November 9, 1979, forester Robert Taylor encountered a large circular-shaped UFO on Dechmont Law—a wooded hill in West Lothian, Scotland. As he approached, two smaller spiked spheres (like sea mines) came out of the UFO and attached themselves to his trousers, dragging him toward the craft. The small probes were making a hissing sound and seemed to be emitting noxious gas. Taylor coughed, choked, and passed out. When he regained consciousness and went home, his wife was so shocked by his disorientated state that she assumed he'd been attacked. She called the police and a criminal investigation was launched. The incident was classed as an assault and Taylor's damaged trousers were examined forensically. The results were inconclusive and the case remains unexplained. One of the skeptical theories put forward was that Taylor had somehow ingested some deadly nightshade (belladonna) and hallucinated the whole affair. It seems highly unlikely that an experienced forester such as Robert Taylor would be unfamiliar with deadly nightshade, let alone that he could somehow inadvertently ingest it.

DELUSIONS, HALLUCINATIONS, AND MASS HYSTERIA

Before completely setting aside the idea of a hallucination, is it possible that some other sort of hallucination—i.e., one not involving drugs—occurred? There are a wide range of psychiatric and psychological conditions (and other more prosaic factors such as physical stress and sleep deprivation) that can result in delusions and hallucinations. Neurologist and bestselling author Dr. Oliver Sacks writes about a range of them in his book *Hallucinations,* in which he briefly mentions alien abductions. Fortunately, we don't need to drill down into the fifth edition of the *Diagnostic and Statistical Manual of Mental Disorders,* published in May 2013, to find an answer here. Rather, we need to ask the more fundamental question as to whether any hallucination or mental disorder could cause several people to experience the same thing.

As it happens, there is one possible candidate: Shared Psychotic Disorder. This new term covers cases where the delusion is shared by two or more people and replaces a multitude of more traditional terms such as "folie à deux," the lesser-known "*folie à plusieurs*" (madness of many), and others. There are also a couple of subcategories to mention here, if only to play devil's advocate. *Folie simultanée* is a situation where two people suffer independently from psychosis but then influence the content of each other's delusions so they become similar or even identical. *Folie imposée* is a situation where a dominant person (sometimes known as a "primary," "inducer," or "principal") develops a delusional belief during a psychotic episode and then imposes it on one or more people who might not otherwise have become delusional. In relation to the first night's events, could John Burroughs and Jim Penniston fall into any of these categories? In relation to the second night, might Halt (perhaps by virtue of his seniority) have been a "primary"?

As ever in the world of psychiatry, there's controversy about a lot of these terms and, in particular, where the boundaries lie. At some point, for example, a shared delusion gets categorized as mass hysteria, which isn't classified as a psychiatric disorder at all. This leaves us with the rather odd position that if a delusion becomes shared by enough people it somehow acquires a "critical mass" that takes those who believe in it out of any psychiatric category!

The wider problem with this theory in relation to Rendlesham is that such psychological conditions would doubtless manifest themselves in all sorts of other ways. In the unlikely event that they weren't picked up in the recruitment and vetting process, they would doubtless have become apparent soon thereafter. There's certainly no way, for example, that such people could fail to be picked up by the Nuclear Weapon PRP.

And in all of the skeptical theories being examined in this chapter we have to consider not just the witness testimony but also evidence such as the radar data and the physical traces at the landing site, including the tree damage and the radiation readings. And while none of these elements are exempt from critical scrutiny, we need to consider how likely it is that all the separate and extraordinary elements of this story—the witness testimony, the radar evidence, the physical evidence at the landing site, the extraordinary way in which the witnesses were treated after the sightings—would have resulted if what happened had some prosaic cause.

POLICE CARS AND PRACTICAL JOKES

There's certainly a culture of pranks and practical jokes in the military. Could some practical joke have gotten out of hand? Is it possible that one (or more) of the witnesses somehow faked the whole thing, over two separate nights and under the noses of the other witnesses? It's difficult to see how this could have been done without it being fairly obvious, and it's difficult to conceive that the guilty party (or parties) would then go along with the story without confessing, to the point where the affair was briefed to the Wing Commander, the UK MoD, the senior USAF officer in the United Kingdom, and then the senior USAF officer in Europe. Even if faking the events on both nights were possible (and it's almost impossible to conceive how this could be done) it seems likely that the guilty party would have confessed at an early stage, nipping the situation in the bud. After all, the reprimand from a shift commander might be a wry smile and a warning not to do it again. The penalty for knowingly making and signing a false statement and/or escalating the issue to the level it actually reached would have been far more serious. And from everything we know about Halt and the other witnesses, they are simply not the sort of people who would have done this—even if such a thing were somehow possible.

Even if we are to believe that on the first night Burroughs and Penniston conspired to concoct a story they would have needed to either include several other participants or make their story convincing enough to fool them. The same would apply if Halt, Englund, Ball, Nevels, and others conspired to pull a separate but related prank on the second night. And while the marks at the landing site could have been faked, the radiation and radar data would be difficult—if not impossible—to manufacture. We are stretching credulity well beyond breaking point here.

Could someone else (i.e., not one of the witnesses) have pulled off a prank of this nature? This would at least get around the objection that the other witnesses would have seen them. It would also get around the point about confessing, as the unknown party would not be putting his or her name to anything and if they weren't seen there would be no adverse comeback, however high the affair was escalated—unless they subsequently decided to confess.

In fact, there's only been one claim of a practical joke that might have had a bearing on this case. The story was published in 2003 in a UK newspaper, the *Daily Mail*. Kevin Conde, a former USAF policeman, claimed that in 1980 he played a prank on the guard at the East Gate. Conde stated: "There was this one guy at the back gate, and he was known as a bit of a problem—he was always seeing things. He had seen lights before and reported them. It always turned out that it was a star or something. So I decided to play a practical joke. I had no idea what I had started by doing this."

It's not clear to whom Conde was referring. He went on: "I drove down the taxiway in my car. I stuck the spotlight on, after sticking red and green lenses on it. I then drove round in circles, in the fog, with the PA loudspeaker going, flashing my lights. It was just a practical joke; we were always playing practical jokes. Then I turned my lights off and drove away." He says that he thought no more of this until years later when he saw mention of the Rendlesham Forest incident on an Internet site for ex-military personnel, put two and two together, and figured that his prank had been the cause of all the trouble.

There are, however, numerous problems with this story. First, by his own admission, Conde cannot recall exactly when he played his prank. He subsequently said that it was during an exercise, but there was no exercise underway at the time of the UFO sightings. Second, Conde claims that there were two other people with him when he carried out his prank but cannot recall the names and nobody has come forward to corroborate his story. Third, the colors of the lights he talks about don't match what was seen, and fourth, the direction is completely wrong.

Even if Conde's claim of having pulled this prank is true, it was a one-off event, not something that spanned two nights. And again, it doesn't explain the radar data or the physical marks and radiation readings at the landing site.

None of the witnesses recall Conde being on duty at the time of the incident. Charles Halt has said, "Just for information, I knew Conde well. I wouldn't have put it past him to claim what he does. He was nowhere to be seen the night I was out. One possible explanation is several nights later he may have opened the back gate while on patrol at Woodbridge and driven down the paved road and displayed his lights to mock the earlier incidents." Burroughs is absolutely clear on this issue: "Kevin Conde was not on duty so he could not have been involved."

There's no independent evidence that this prank even took place. If it did, it's just possible, as Halt suggests, that it took place sometime after the UFO sightings, in an attempt to make people think the UFO had returned.

TRACTORS, TRUCKS, AND MANURE

In September 2009 a new claim emerged concerning the 1980 UFO sightings. Peter Turtill, from Ipswich, claimed that he had been driving a truck filled with fertilizer down a road close to the twin bases, having picked it up from a friend to whom he'd lent it. He went on to say that the truck broke down and at this point he realized that the fertilizer was stolen. He and a friend towed it off the road and into Rendlesham Forest, where they allegedly set fire to it in an attempt to destroy the evidence. Turtill says that a combination of the metal from the truck and the fertilizer created some oddly colored flames and believes that this burning truck, being towed through the forest (before being towed back to the road to avoid further spooking some soldiers with guns, as the story goes on!), was what caused the UFO sightings. Turtill summed up the situation thus: "There was no real fuss at the time and it was only later people started saying it was aliens and this story spiraled." He even claimed that the burned-out vehicle remained in the forest for twenty years afterwards until it was finally removed. We have been unable to verify this or locate any police records or contemporaneous newspaper story relating to the alleged theft.

To say that all this is an exceptionally unlikely chain of events is an understatement. As it transpires, Turtill is a well-known local character who campaigns on issues such as rights-of-way/access to the countryside and who once stood for election as an independent Member of Parliament— but polled only ninety-three votes. We include his claim only for the sake of completeness and to show that such is the media fascination with the Rendlesham Forest incident that these completely unsubstantiated claims still managed to make at least three UK national newspapers.

Entering the name "Peter Turtill" into any search engine will throw up information that certainly paints a picture of a controversial and colorful character. This doesn't automatically invalidate his Rendlesham claim, of course, but it should certainly raise a few red flags. The problem with Rendlesham is that the case is so newsworthy, any tall tale or speculative remark can become a viable theory in the minds of some people.

A similarly unlikely story had surfaced a few years earlier when a skeptical ufologist made a throwaway remark about how the colors and configuration of the lights on the UFO seen on the first night sounded similar to the array of lights carried at night on a tractor. It wasn't even clear if the comment was serious or tongue-in-cheek, but next thing, UFO discussions forums were abuzz with the "theory."

In evidential terms, the burning truckload of fertilizer story—like the police car tale—hangs on unsubstantiated claims from a single individual. There's no independent evidence that such a crime or prank ever occurred, let alone evidence that such events (even if they did take place) had anything to do with the Rendlesham Forest incident. Any intelligence analyst will caution against hanging a decision on a "single, unverified source."

METEORS AND FIREBALLS

Astronomical events such as meteors and fireballs have been known to generate UFO reports from people not used to seeing such things. Meteors are caused when small rocks or dust particles burn up in the Earth's atmosphere. They streak across the sky in a straight line and are generally visible only for a second or two. They are a relatively common sight for anyone accustomed to looking at the night sky, though they are more difficult to spot from cities, in view of light pollution. A fireball is essentially just a very bright meteorite, and indeed the International Astronomical Union defines a fireball as "a meteor brighter than any of the planets." The term "bolide" is also used on occasion to describe a particularly bright fireball.

The British Astronomical Association keeps records of such activity, and their Meteor Section's newsletter for the time period in question reports that three fireballs were observed on the night of December 25/26 1980: one at around 5:20 pm, another at around 7:20 pm, and the final one at around 2:50 am. This final one overlaps with the time period when Burroughs, Penniston, and others were seeing the UFO. However, the critical point here is that according to the British Astronomical Association, each of these fireballs was visible for a time period of only a few seconds.

A ROCKET RE-ENTRY

At around 9:07 pm or 9:08 pm on the night of December 25, 1980, a considerably brighter and longer-lasting aerial phenomenon would have been visible in the skies of northwest Europe. It was the British Astronomical Association that first documented this, and indeed the first reports suggested that what had been seen was a particularly bright fireball. However, research conducted by John Mason and Howard Miles of the British

Astronomical Association determined that what had actually been seen was part of a Soviet rocket re-entering and burning up in the Earth's atmosphere.

The debris concerned was identified as part of the rocket that had launched (in 1975) the Cosmos 749 satellite. Mason and Miles collated reports from Morocco, Spain, Portugal, France, and the United Kingdom. In all, around one hundred reports were received, but it's highly likely that despite extensive appeals in the media and in astronomical publications, there were many other witnesses who never came forward. Mason and Miles plotted the likely track of the debris as it broke up, and it seems that it might just have been visible from parts of the county of Suffolk, though the best estimate is that the debris had either burned up or lost its luminosity by this point.

To observers, the sight would be somewhat like a bright firework display, at high altitude, with clusters of different-colored objects moving slowly, in a straight line. Again, however, this simply doesn't fit what the Rendlesham Forest incident witnesses saw. About the only UFO connection is the fact that something unusual was seen in the sky. But of course, the first sighting was of something on the ground, not in the sky—and the rocket re-entry took place some hours before the initial sighting of the lights on the ground in the forest. Even if the re-entry of part of the Cosmos 749 rocket may just have been visible from the area, it took place before the start of the Rendlesham Forest incident and it seems unlikely the rocket played a part in the events.

THE APOLLO COMMAND MODULE

One of the units based at Bentwaters/Woodbridge was the 67th ARRS (Aerospace Rescue and Recovery Service)—now redesignated as the 67th Special Operations Squadron. Its primary mission is described as "providing worldwide clandestine aerial refueling of special operations helicopters," with a secondary capability that includes "infiltration, exfiltration, and resupply of special operations forces by airdrop or airland tactics." One other specialist mission that this unit trained for was recovery operations for the Apollo and Skylab programs. Part of the training for this mission included an object known as Boilerplate, Command Module, Apollo, #1206. This object weighed nine thousand pounds and resembled—externally— an Apollo command module.

Though the Apollo program ended in 1972 and the Skylab program in 1979, the 67th ARRS certainly still possessed the boilerplate at the time of the Rendlesham Forest incident. Could the roughly triangular object have been accidentally dropped in the forest while being transported by helicopter as a slung load? Alternatively, could the boilerplate have been placed there as a prank?

As with some of the previously discussed theories, what we have here is one factor (the size and shape of the boilerplate) that matches in some respects a particular part of the story—Burroughs and Penniston encountering a triangular craft in the forest. But look beyond this single match and the theory soon breaks down. If there was a helicopter flying that night (and the consensus is that there was no flying activity at all at the time of the incident), the witnesses would have seen and heard it. Even if we eliminate the helicopter and imagine the boilerplate being placed in the clearing in advance, as a practical joke, the facts simply can't be made to fit. While the boilerplate was heavy enough to leave an impression on the ground, the underside had no features capable of having produced the three indentations in the triangular pattern. Even if the holes were made separately, the theory completely ignores the movement of the craft Burroughs and Penniston saw ("speed—impossible" was how Penniston described the sudden acceleration in his police notebook) and is totally at odds with what Halt and his men saw on the second night, with the UFO firing beams of light down at the ground.

Finally, there is not a single document or eyewitness that would support the theory.

A LIGHTHOUSE AND A LIGHTSHIP

Another theory is that the witnesses misidentified the nearby Orfordness lighthouse (and/or the Shipwash lightvessel). This idea was first proposed by local forester Vince Thurkettle.

The first question that arises is the extent to which the witnesses were or weren't familiar with the lighthouse. While it was a well-known local feature, some of the more recent arrivals may not necessarily have known about it, though personnel who had been there for any length of time would doubtless have encountered it.

A more fundamental objection is the fact that the lighthouse simply isn't visible from most of the locations from which the witnesses saw the UFO. That said, it's difficult to be definitive about this, because Rendlesham Forest was hit extremely hard by the Great Storm of 1987 (when near-hurricane-force winds hit the United Kingdom on October 15/16, 1987, killing at least twenty-two people) and a large proportion of the trees were flattened. The forest was replanted, and while the topography hasn't changed, the height and layout of the trees have, making comparisons with 1980 very difficult.

The key point about the lighthouse is that it crops up in four of the five witness statements cited in chapter 7. The relevant extracts are as follows:

Buran: SSgt Penniston reported getting near the "object" and then all of a sudden said they had gone past it and were looking at a marker beacon that was in the same direction as the other lights. I asked him if he could have been mistaken, to which Penniston replied that had I seen the other lights I would know the difference.

Chandler: He eventually arrived at a "beacon light," however, he stated that this was not the light or lights he had originally observed.

Burroughs: Once we reached the farmer's house we could see a beacon going around, so we went toward it. We followed it for about 2 miles before we could see it was coming from a lighthouse.

Cabansag: While we walked, each one of us would see the lights. Blue, red, white, and yellow. The beacon light turned out to be the yellow light. We would see them periodically, but not in a specific pattern. As we approached, the lights would seem to be at the edge of the forest. We were about 100 meters from the edge of the forest when I saw a quick movement, it look visible for a moment. It looked like it spun left a quarter of a turn, then it was gone. I advised SSgt Penniston and A1C Burroughs. We advised CSC and proceeded in extreme caution. When we got about 75–50 meters, MSgt Chandler/Flight Chief, was on the scene. CSC was not reading our transmissions very well, so we used MSgt Chandler as a go-between. He remained back at out vehicle. As we entered the forest, the blue and red lights were not visible anymore. Only the beacon light was still blinking. We figured the lights were coming from past the forest, since nothing was visible when we passed through the woody forest. We would see a glowing near the beacon light, but as we got closer we found it to be a lit up farm house. After we had passed through the forest, we thought it had to be an aircraft accident. So did CSC as well. But we ran and walked a good 2 miles past out the vehicle, until we got to a vantage point where we could determine that what we were chasing was only a beacon light off in the distance. Our route through the forest and field was a direct one, straight towards the light. We informed CSC that the light beacon was farther than we thought, so CSC terminated our investigation.

There are, as we have seen, serious problems with these five witness statements, with some of the witnesses claiming that these were "sanitized" versions of events that they were made to sign. Might it be that the lighthouse was selected as a convenient cover story? Cabansag's brief statement manages to use the phrase "beacon light" or "light beacon" an extraordinary five times. It's similarly odd that the word "beacon" crops up in four of the statements, though this could well be a consequence of the witnesses discussing their shared experience and picking up on one another's terminology. But it's also odd that the word "lighthouse" only appears once.

Though we mention the possibility of the lighthouse being selected as a cover story, this doesn't really work, as the five statements—whatever their true provenance—mention the light beacon but go on to make it clear that while this was visible at times, it was a separate light from the unusual lights that they were following.

As with most of the theories discussed to date, an idea that starts out as being superficially attractive in terms of explaining a small part of the case soon breaks down if one tries to shoehorn in the wider facts of the two nights' encounters. The lighthouse cannot possibly explain Charles Halt's sighting of a UFO firing beams down at him and his men from directly overhead ("Lighthouses don't fly" was his terse response to the suggestion—a comment echoed by Burroughs, who stated, "The lighthouse does not land, or hover on the ground and take off at a high rate of speed. Remember, we got close to whatever it was") or hovering over the base at Woodbridge. Indeed, Halt has gone on the record to say that at one point he calculated the bearing of the UFO by comparing it to the position of the lighthouse—illustrating the point that they were two separate objects. And yet again, this theory fails to take account of the radar evidence or the physical trace evidence at the landing site.

While not the most compelling of the skeptical theories (a "top three" follow), the lighthouse theory has become the one most favored by TV documentaries that have covered this case. There are four reasons for this. First, the theory has—in Vince Thurkettle—an informed, local proponent. Second, it sounds so absurd and is so insulting to the witnesses and the believers that it provokes a strong reaction. Third, it's a very visual theory that lends itself to being re-created and can be tested. All these three factors make for good, dramatic TV. The fourth reason is that the MoD gave the theory a subtle push, as part of a wider campaign to "spin" the Rendlesham story by downplaying the events.

Another nail in the lighthouse coffin was provided by the lighthouse keeper himself, in a conversation with author and investigative journalist Georgina Bruni, when he said, "The skeptics have been pestering me in an attempt to get to support their theory. I cannot do it. I know what my lighthouse looked like from the forest. I have seen it in all weathers. It just could not do what those airmen and local people describe the UFO as doing."

In fact, where the lighthouse is visible from the forest at all, it appears in the distance, as a tiny pinprick of light on the distant horizon. This is not sufficient, however, to completely eliminate the lighthouse, as it was brighter in 1980 than it was in later years. In addition, while the lighthouse was fitted with a shade that prevented it illuminating the town of Orford, this wasn't positioned in such a way as to render the glare invisible from some parts of the forest. Vince Thurkettle has made it clear that there were plenty of locations from which it was visible.

As a final observation on this theory, I was filming with Vince Thurkettle in Rendlesham Forest in November 2010, along with Charles Halt. We discussed the matter at great length, and while Thurkettle still maintained that it was possible that the lighthouse could have played a small part in events, he did not believe it could fool people over two nights or for so long a period of time. So while some UFO skeptics have appropriated the lighthouse theory as their own, its original proponent has effectively ruled it out. Discussing his theory with Georgina Bruni, Thurkettle expressed irritation with UFO skeptics who had hijacked his speculative musings on the lighthouse and used it as their own definitive explanation: "They take a cluster of facts and only pick up on those that suit the situation."

GUARD FORCE TESTS, GHOST GUNS, AND MIND CONTROL

Is it possible that the Rendlesham Forest incident was attributable to a highly unusual guard force test? There are several versions of this theory, but the general thrust of the idea is that the various UFO sightings were deliberately staged by some elements of the chain of command to see how the guard force would react. The thinking is that if exercise planners could devise a scenario where the guard force would react (or fail to react) in such a way as to leave critical systems vulnerable, a hostile power could do the same. Thus, the purpose of such a test would be to see if a weakness exists and, if so, to plug the gap. Suggestions as to how such an exotic test could have been accomplished include hallucinogenic drugs, mind control, and holographic technology. The latter idea, in particular, has been the subject of some considerable speculation in not just the UFO community but also the wider conspiracy theory community.

Holographic technology exists and it's known that various defense contractors, among others, have experimented with this. Phrases such as "ghost gun" have been used to describe the equipment necessary to create images. But there are two problems here. First, such programs tend to be classified and the extent to which this technology exists in any usable form is unknown. Second, the waters have been muddied by the existence of a conspiracy theory known as "Project Blue Beam," which suggests that holograms/projections (perhaps combined with more conventional Hollywood special effects) of things such as an alien invasion or the Second Coming might be used to create mass panic as part of a "false flag" event that would allow the powers that be (e.g., the Illuminati) to create a "New World Order." This is, to say the least, somewhat far-fetched.

It's undeniable that the military wants to know how its guard force is going to react in certain circumstances, especially at the most important military facilities. Realistic training underpins this, as it underpins much in the military, and regular exercises are held to gauge the response to various different scenarios. Is it too much of a stretch of the imagination to suggest that guards at Bentwaters/Woodbridge were hoodwinked into thinking they faced some unknown—perhaps extraterrestrial—phenomenon, to see how they'd react? But is there any real evidence, aside from speculation, to support such a theory?

A US document titled *The Role of Behavioral Science in Physical Security—Proceedings of the Fifth Annual Symposium, June 11–12, 1980* has put this theory in the spotlight in the UFO community. The fact that this was a mere six months before the events in Rendlesham Forest has added additional poignancy. The symposium was organized by an official in the Nuclear Safety Directorate of the Defense Nuclear Agency. Enhancement of nuclear weapons storage was a key aim. Sponsors included the Law Enforcement Standards Laboratory. A paper titled "Security System Operational Recording and Analysis (SSOPRA)" described a plan to evaluate security at a Weapons Storage Area at an Air Force base, using a variety of techniques including "active procedures" to test the system. It is clear why many people believe this is relevant to Bentwaters/Woodbridge.

The problem is that none of this suggests that the "active procedures" should be exotic. Indeed, the paper states that "initiating events will be of the types commonly employed by evaluation teams." And neither is there any evidence that Bentwaters/Woodbridge was selected for the SSOPRA trial. Moreover, when the paper was considered, a discussion took place about the possibility of "black hat assaults as a test of the system." The response pretty much rules this out as a potential explanation for the Rendlesham Forest incident: ". . . everybody early in the game knows it is an exercise . . . what you cannot do as far as I can see at the present time is to actually initiate an event that will result in the call out of a fire team, for example, without the whole system knowing it is an exercise. That simply is beyond the scope of what we might be able to do."

Notwithstanding this, there are a few accounts of UK Special Forces being called upon to launch unannounced incursions into various military installations, including Bentwaters/Woodbridge. Such incursions—as one might expect, with elite troops such as the SAS (Special Air Service)—were often successful, resulting in huge embarrassment for those charged with security at the locations concerned.

If "guinea pigs" were needed for a test to see how security personnel would respond to an exotic scenario, conducting the test in the United Kingdom would have been an unnecessary complication and using a base in the United States would have been far preferable. But even if the Rendlesham Forest incident had been attributable to a trial along these lines, we strongly suspect that someone would have taken those concerned aside afterwards, told them that they'd been taking part in a trial, said that the details were classified, and instructed them not to say anything. As loyal military personnel, they would doubtless have complied. There might have been some private chat, but we very much doubt anyone would have gone public in the way that they did.

A NUCLEAR ACCIDENT OR INCIDENT

The US DoD has a list of codes to cover nuclear weapons accidents or incidents. These are, with the most serious first, "Broken Arrow," "Bent Spear," "Empty Quiver," and "Faded Giant." A previous example of a "Broken Arrow" incident at a US base in the United Kingdom had reportedly occurred on July 27, 1956, when a B-47 bomber crashed and exploded into a nuclear weapons storage facility at the Lakenheath air base. According to various reports in the public domain, the initial report to the Commander of US Strategic Air Command ended with the statement: "Preliminary exam by bomb disposal officer says a miracle that one Mark Six with exposed detonators sheared didn't go [off]." Could the UFO story have been a deliberately concocted piece of fiction designed to distract attention from another "Broken Arrow" type of event? Might this, for example, explain the radiation readings taken at the landing site?

Again, we are constrained here by the US and UK government policy neither to confirm nor deny the presence of nuclear weapons at any particular time or place. However, common sense dictates that the US government would either admit to the incident or decline to comment. Concocting a bogus cover story seems highly unlikely, but even if one accepted the possibility, a UFO story would be the worst possible idea, given media and public fascination with the subject. Such a story would invariably result in the last thing any cover story should generate: attention and publicity.

A related theory is that the event was in some way related (again, perhaps as a cover story) to a nuclear incident (e.g., a leak) at the nearby Sizewell A or Sizewell B nuclear power station. Again, however, there's no witness testimony or documentary evidence to support such a theory.

SECRET PROTOTYPE AIRCRAFT AND DRONES

At any given time there will be various aircraft and drones being developed, test flown, and even deployed operationally but about which no public announcement will have been made. In some cases such technologies are not publicly declared for perhaps ten or fifteen years. The F-117 stealth fighter and the B-2 stealth bomber are two good examples. The stealth fighter was publicly declared in 1989 but had been flying for many years prior to that.

Could the Rendlesham Forest incident have been caused by the witnesses misidentifying some new spy plane or—more likely, given the small size of the object Burroughs and Penniston encountered in the clearing—drone? And in such circumstances, is it possible that the witnesses (and even perhaps the chain of command) never discovered the explanation or, if they did, deployed a UFO-related cover story to hide the truth? Intriguingly, there is some historical precedent for this. The CIA article "CIA's Role in the Study of UFOs, 1947–90: A Die-Hard Issue" makes it clear that it suited the CIA and the USAF when sightings of the U-2 spy plane or the SR-71 reconnaissance aircraft were reported as UFOs, as foreign military intelligence analysts were less likely to pay attention to UFO stories than stories about strange aircraft. Might something similar have happened in Rendlesham Forest, notwithstanding our earlier point about the apparent unattractiveness of a UFO cover story, in view of public and media interest in UFOs?

At first glance, if a secret aircraft or drone *had* been involved and if rumors of this were circulating among USAF personnel and the local population, it might have made sense to launch a counter-intelligence operation and "rebrand" the event as a UFO sighting. But there's no suggestion that any such rumors had started, so there was no need for a cover story.

Furthermore, even if there had been a secret aircraft or drone involved, a cover story would have been unnecessary; military witnesses would simply have been told that they'd seen something classified (they are unlikely to have been given any further details, in line with the need-to-know principle) and ordered not to discuss what they'd seen. It is highly unlikely that they would have disobeyed such an order.

Moreover, while a UFO cover story may have been effective in the early days of the U-2 and the SR-71 programs, as we mentioned in respect of the theory concerning a test of the guard force, a UFO cover story in the United Kingdom in 1980 would have done the one single thing that any effective cover story must not do: attract media and public attention.

Furthermore, it's highly unlikely that Halt would have sent his famous memo to the MoD if he or anyone else in the chain of command had known that a secret aircraft or drone was involved. Someone in US intelligence would have had a quiet word with someone in UK intelligence and the whole business would have been forgotten.

There is, however, one variation on this theory that might make a little more sense. What if the aircraft or drone had been Soviet? Might the strange symbols that Penniston saw on the side of the craft have been Cyrillic script? This was certainly a time of high international tension, related to the rise of the Polish trade union Solidarity and the Bentwaters and Woodbridge bases were two of the most important military facilities in the United Kingdom—and indeed in NATO. Did a Soviet drone crash? Was it shot down, inadvertently or deliberately? This latter action could have started World War Three, and in that scenario might cooler heads on both sides have prevailed and buried the incident?

Critically, all three governments involved would have had a good reason for saying nothing or even developing a cover story. The Soviets would not admit to having been conducting espionage activity of this sort, while the United States and the United Kingdom would have been horrified that the United Kingdom's Air Defence Region had been penetrated with such ease and would not wish any knowledge of such a security breach to circulate among junior military personnel (for morale reasons) or to become public knowledge. Perhaps a small group of Soviet, American, and British military intelligence personnel conspired to keep this event from their political masters and mistresses, aware that this could quickly escalate matters to a dangerous level. Such a cover-up would probably have been the right thing to do.

All this, however, is pure speculation. There's not a single shred of documentary evidence or personal testimony that would support this interpretation, and it's offered up only as a hypothetical scenario that, frankly, makes more sense than most of the theories bandied around by skeptics in the UFO community.

Now that we have run through the possible skeptical theories and found them, at best, unsupported by any evidence and, at worst, demonstrably false, it's time to consider other more exotic theories. Follow us now on a trip into the real-life Twilight Zone.

10. EXOTIC THEORIES

Having considered but pretty much eliminated the skeptical theories about the Rendlesham Forest UFO incident, we must now at least consider some more exotic possibilities. In this chapter, we're going to look at three such possibilities: extraterrestrial visitation, interaction with currently hidden dimensions and parallel universes, and time travel. Why should we even *consider* concepts that many regard as science fiction? There are several reasons. First, vast numbers of people believe these things. Admittedly, lots of people believe things that aren't necessarily true, but any widely held belief is interesting in and of itself, even if only from a psychological perspective. Second, several of the Rendlesham witnesses support these theories, and as they were the ones who went through these events, it seems entirely proper to highlight their views and see why they believe in these explanations. Third, it's as well to say "never say never" with some things that sound highly unlikely but might have a big payoff if they turned out to be true. It's what the business community calls low probability/high impact (or sometimes low probability/high consequence). Indeed, it's this very mind-set that explains why governments look (as they indisputably do) at things like UFOs and remote viewing (a more respectable-sounding term for psychic spying) in the first place. Finally, we do so for the sake of completeness. Any book on this incident needs to examine all the theories that are either tenable and/or widely held. Many of the skeptical theories turned out to be somewhere on the spectrum between highly unlikely and impossible, but it was unfair to omit them and thus miss the chance to subject them to some degree of scrutiny. So it is with the more exotic theories. Some people may think they're science fiction, but let's at least take a look, if only to satisfy ourselves on this point.

We intend to look at the three aforementioned exotic theories in a slightly different way from the way in which we examined the skeptical theories in the last chapter. Because of the controversial nature of the concepts involved, we intend to take a step back and ask not whether things like extraterrestrial visitation could explain the Rendlesham Forest incident but whether such things are even possible in the first place. Believers may say this is a little unfair and sets the "evidence bar" a little higher, but it's a far easier way of attacking the problem, because if extraterrestrial visitation is impossible *in and of itself* clearly we can eliminate the theory straightaway. And in any case, "extraordinary claims require extraordinary evidence," as the famous quote from the cosmologist and author Carl Sagan goes.

EXTRATERRESTRIAL VISITATION

The term "UFO" stands for "Unidentified Flying Object" but in popular culture has become synonymous with alien spacecraft. This leads to nonsensical questions such as "Do you believe in UFOs?" where the answer "no" is patently absurd if the term is used correctly. The fact remains, huge numbers of people not only believe in the existence of extraterrestrial life but also think the Earth is being visited by intelligent extraterrestrials and that elements within the US government (and perhaps other governments) are aware of this but are actively conspiring to cover up this truth. How likely is any of this?

Before addressing the question of UFOs, we need to take a step back and ask some more fundamental questions. Are there any scientific objections to the existence of life elsewhere in the universe and, if not, is it scientifically possible that any of this life could be visiting us?

One often hears phrases along the lines of "it would be arrogant and ridiculous to assume that in this infinite universe we're the only life."

While this argument undoubtedly has a certain attractiveness, it's not one that uses science to come to its conclusion and thus, we must set it aside.

From a purely scientific point of view, observational data suggests that the laws of physics and the laws of chemistry are constant in the universe. This being the case, given the sheer number of stars and (as we are now in the process of discovering) planetary systems in the universe, it is a reasonable hypothesis that the factors that gave rise to life here on Earth should arise many times.

There are however, problems with this hypothesis. Even if we set aside religious beliefs about Creation, there's no scientific consensus on exactly *how* life arose on Earth or even on the more fundamental question of what constitutes life—on this latter point see, for example, the scientific debate over whether or not viruses constitute life or the more speculative debate over artificial intelligence, which soon segues from science to philosophy. Moreover, there are some who believe that life may not have originated on Earth but might have been seeded here—not in the way that we see in sci-fi movies like *Prometheus* or UFO TV shows like *Ancient Aliens* but through the mechanism of organic material arriving through cometary impact.

Even if life has arisen many times in the cosmos, is any of it intelligent, or is there something unique about complex life and—perhaps—intelligence? The so-called Rare Earth hypothesis (postulated by astronomer Donald E. Brownlee and paleontologist Peter Ward) suggests that while microbial life in the universe might be common, advanced life (i.e., animals) is rare. They do not, however, quantify the term "rare"! Another counter-argument to the idea that intelligence is unique is the fact that in evolutionary terms being smart seems to be a pretty good survival strategy.

The Rare Earth hypothesis is seen as a response to the views of scientists such as Carl Sagan and Frank Drake, who have expressed the view that intelligent life is likely to be widespread in the universe. In 1961 astronomer Frank Drake devised an equation (now known as the Drake Equation) that attempted to calculate the number of detectable extraterrestrial civilizations in our own galaxy (the Milky Way) by calculating the values of a number of relevant variables. To put things in perspective, the Milky Way is estimated to contain somewhere between 100 billion and 400 billion stars. Estimates for the number of galaxies in the universe vary between 100 billion and 200 billion. The Drake Equation is as follows:

$$N = R^* \times fp \times ne \times fl \times fi \times fc \times L$$

In this equation:

N = the number of civilizations in the Milky Way galaxy whose electromagnetic emissions are detectable

R^* = the rate of formation of stars suitable for the development of intelligent life

fp = the fraction of those stars with planetary systems

ne = the number of planets, per solar system, with an environment suitable for life

fl = the fraction of suitable planets on which life actually appears

fi = the fraction of life-bearing planets on which intelligent life emerged

fc = the fraction of civilizations that developed a technology that releases detectable signs of their existence into space

L = the length of time such civilizations release detectable signals into space

The problem with this is that while some values can be estimated with some degree of accuracy, most can only be guesses. At the front end of the equation, for example, it seems reasonable to make some calculations about star formation, based on observational data. And while a few years ago the existence of any planetary systems aside from our own was speculation, by the end of November 2013, 1,039 extrasolar planets had been discovered. A November 2013 study based on data from NASA's Kepler space telescope estimated that one in five sun-like stars in our galaxy would have an Earth-like planet in its habitable zone, meaning that a value for fp can be better estimated than was previously the case. But the last four values can really only be guesses in the absence of the discovery of extraterrestrial life. Thus, while Frank Drake estimates that there are around ten thousand detectable civilizations in our galaxy, critics have said that the constituent parts of the equation can vary so much that the actual figure of N could be anywhere between zero and several billion!

How close are we getting to discovering the existence of extraterrestrial life? We've had a couple of close calls already. In 1996 NASA scientists announced that a meteorite that had originated on Mars (ALH84001) seemed to contain fossilized traces of bacteria. The data have since been disputed and the issue remains unresolved. More recently, in 2009, other NASA scientists expressed the belief that methane on Mars might be a by-product of microbial life under the ground. In a sense, this doesn't help us. We know that through the impact of meteors material can be thrown up into the Martian atmosphere, escape from the planet's weak gravitational field, and arrive on Earth—precisely what happened with ALH84001. This being the case, how we can be certain that any life found on Mars didn't originate on Earth? Indeed, how can we be sure that life on Earth didn't originate on Mars? This latter idea was the headline grabber at the prestigious Goldschmidt meeting held in Florence in August 2013, where Professor Steven Benner suggested that conditions for the development of life in the early solar system were actually more favorable on Mars than Earth and that life likely developed on the Red Planet before being transported to Earth via meteorites.

The search for extraterrestrial life continues and encompasses a number of different locations and methodologies. Agencies such as NASA would like to launch missions to find life (probably microbial life) on Mars or Europa. Other scientists are focusing on the aforementioned extrasolar planets and in particular on the search for Earth-like planets orbiting sunlike stars. The first super-Earths have been discovered and there's a growing sense that it's only a matter of time before we find what science dubs Earth 2.0. The longer-term goal will be to use the next generation of space telescopes to undertake spectral analysis of the atmospheres of exoplanets, looking for oxygen, ozone, water, and other potential chemical indicators of life. Another search strategy involves the use of radio telescopes to search for a signal from other civilizations. Frank Drake was a founding father of this work, dubbed SETI (Search for Extraterrestrial Intelligence). In 2024, a radio telescope orders of magnitude bigger and more powerful than any existing array will become fully operational. Scientists claim that the Square Kilometer Array will be so powerful, it will be able to detect an airport radar at a distance of fifty light-years. If there are detectable civilizations in our small corner of the galaxy, the SKA should be able to find them—provided SETI scientists get sufficient telescope time on the array!

How seriously is any of this taken? It's clearly science, but where does it lie on the spectrum—mainstream or maverick? A few years ago the answer would have been the latter. However, in 2010 the scientific establishment sent a strong message that the search for alien life was now respectable. The prestigious Royal Society (founded in 1660, its members have included legendary figures such as Isaac Newton, Charles Darwin, Albert Einstein, and Stephen Hawking) hosted two discussion meetings, titled, respectively, "The Detection of Extra-terrestrial Life and the Consequences for Science and Society" and "Towards a Scientific and Societal Agenda on Extra-terrestrial Life." These two meetings considered not just the progress of the search for alien life but also more speculative questions such as whether or not people would panic and what effect the discovery would have on religious belief.

In another sign that the question of alien life is no longer taboo to mainstream science, NASA now has a formal astrobiology program. The mission statement on the relevant part of NASA's Web site begins as follows:

> Astrobiology is the study of the origin, evolution, distribution, and future of life in the universe. This multidisciplinary field encompasses the search for habitable environments in our Solar System and habitable planets outside our Solar System, the search for evidence of prebiotic chemistry and life on Mars and other bodies in our Solar System, laboratory and field research into the origins and early evolution of life on Earth, and studies of the potential for life to adapt to challenges on Earth and in space.
>
> NASA's Astrobiology Program addresses three fundamental questions: How does life begin and evolve? Is there life beyond Earth and, if so, how can we detect it? What is the future of life on Earth and in the universe? In striving to answer these questions and improve understanding of biological, planetary and cosmic phenomena and relationships among them, experts in astronomy and astrophysics, Earth and planetary sciences, microbiology and evolutionary biology, cosmo-

chemistry, and other relevant disciplines are participating in astro-
biology research and helping to advance the enterprise of space
exploration.

Before leaving aside the question of whether or not there's intelligent life elsewhere in the universe, we should mention the Fermi Paradox. Named after the physicist Enrico Fermi, the paradox is usually written as "Where is everybody?" Fermi's point was that if—as scientists such as Frank Drake suggest—the universe is teeming with life, why has Earth not been visited and why is there no other observable or detectable evidence of extraterrestrial life? The numerous responses to the Fermi Paradox include the suggestion that evidence exists but is being covered up, along with theories that other civilizations arise but are wiped out by the sorts of calamities that might yet wipe us out: nuclear war, climate change, plague, the rise of artificial intelligence, asteroid strikes, gamma ray bursts, et cetera. Others have speculated that the Earth is being deliberately avoided, perhaps by advanced civilizations that choose to let humanity develop naturally, without external interference. This idea—broadly similar to the "Prime Directive" in the *Star Trek* franchise—is known as the Zoo Hypothesis. A similar theory—the Planetarium Hypothesis—suggests that we're being deliberately isolated by advanced civilizations and that the observable universe is actually a simulation, designed to look empty. There are numerous other theories and counter-arguments, but they are largely beyond the scope of this book.

As it happens, there's one way that the question of whether or not alien life exists might be resolved without our actually needing to find it! Many scientists will tell you that life arose almost as quickly as was possible on Planet Earth (the Earth is around 4.5 billion years old and life seems to have arisen about 3.5 billion years ago, only shortly after massive planetary impacts ceased), suggesting life develops easily, meaning it should be common in the cosmos. Yet so far as we know, all life on Earth can be traced back to a single "Universal Common Ancestor." So if life arises so easily, why—apparently—has it arisen only once? Why not several times? The shadow biosphere (sometimes dubbed Genesis 2) is the name given to a theoretical second strand of life on Earth. Because if life arose not once but twice (or multiple times) on this single planet, it would indeed suggest that life is not some cosmic miracle but arises easily. The clear and logical follow-on would be that the universe is teeming with life.

The search for a shadow biosphere is arguably much easier than the search for extraterrestrial life, and there has already been some controversy about this. On December 2, 2010, NASA held a press conference with biochemist Felisa Wolfe-Simon. Wolfe-Simon has been doing research at Mono Lake, California, and the press conference was used to announce the discovery of microbes that can apparently use arsenic instead of phosphorus to build their DNA. This fundamentally changes how we define life, and even though this wasn't announced as proof of the shadow biosphere, it came tantalizingly close. Wolfe-Simon's results have since been challenged and the debate is ongoing.

In relation to the Rendlesham Forest incident, the question of whether or not extraterrestrial life exists is fundamental, because this has long been a favored theory of many people in relation to the events. With some people, this is a question of passionate belief. With others, it's because this is the one UFO case that seems to defy any conventional explanation. None of the skeptical theories stand up to detailed scrutiny, and now that Roswell is so far in the past Rendlesham is the case that's become the standard-bearer for what the UFO community calls the ETH (extraterrestrial hypothesis). However, there's a further hurdle that needs to be cleared if we're to seriously entertain the idea that the Rendlesham Forest incident is attributable to extraterrestrial visitation. Because even if we accept that there may be other civilizations out there, if they can't get here it's case closed.

The clear scientific consensus is that faster-than-light travel is impossible. Under Einstein's Special Theory of Relativity, an infinite amount of energy would be required (a clear impossibility) to accelerate a particle to the speed of light. It's worth making a few points about this. First, there are plenty of theoretical physicists who—while they accept that an object can't be accelerated to, or past, light speed—think that there might be viable workarounds, including using wormholes and warping space-time. Second, there's nothing in the Special Theory of Relativity that precludes the existence of particles that always travel faster than light. However, while the existence of such particles (called tachyons) has been postulated, no significant evidence has been found that would support their existence. In any case, it's difficult to envisage a way in which the existence of such particles would help in the construction of a spaceship capable of interstellar travel.

The good news here is that even if faster-than-light travel is impossible, interstellar travel is not. Theoretical physicists have speculated about the possible development of spacecraft powered by propulsion systems based on nuclear fission or matter-antimatter annihilation. None of this would offend the laws of physics (and even if it did, we should remember that our understanding of the laws of physics is constantly changing and evolving). Writing in the February 1999 issue of *Scientific American*, Stephanie D. Leifer, manager of advanced-propulsion concepts at the Jet Propulsion Laboratory in Pasadena, wrote about such technologies. She said about them: "For these will be the technologies with which humanity will finally and truly reach for the stars." More recently, a joint NASA/DARPA (Defense Advanced Research Projects Agency) initiative called the 100 Year Starship Study led to the creation of "a viable and sustainable non-governmental organization for persistent, long-term, private-sector investment into the myriad of disciplines needed to make long-distance space travel possible." Their mission statement reads as follows:

> 100 Year Starship will pursue national and global initiatives, and galvanize public and private leadership and grassroots support, to assure that human travel beyond our solar system and to another star can be a reality within the next century. 100 Year Starship will unreservedly dedicate itself to identifying and pushing the radical leaps in knowledge and technology needed to achieve interstellar flight while pioneering and transforming breakthrough applications to enhance the quality of life on earth. We will actively include the broadest swath of people in understanding, shaping, and implementing our mission.

A propulsion system capable of even a smallish fraction of the speed of light could put the stars within our reach—and put us within reach of any spacefaring civilizations in our local part of the galaxy.

For purists who think this goes too far (or for engineers, who rightly point out that such speculation from theoretical physicists will only come to fruition if engineers can actually build something), let's limit ourselves to the technology we already have. Our Pioneer 10 spacecraft will reach the red giant Alderbaran (sixty-eight light-years away) in about 2 million years from now. Pioneer 11 will pass close to one of the stars in the constellation of Aquila in around 4 million years. Pioneer 10 and Pioneer 11 carry plaques with messages for any intelligent extraterrestrials who may one day encounter them. Part of the messages on these plaques used binary code to give the location of Planet Earth.

The Voyager 1 spacecraft isn't heading for a particular star, but in around forty thousand years will pass within 1.6 light-years of a star that lies some 17.6 light-years from Earth. The Voyager 2 spacecraft isn't heading for a particular star, either, but will pass within about 4 light-years of Sirius in around 296,000 years. Both Voyager craft carry messages for any extraterrestrials who might find them. The point of all this is to show that given enough time, our own technology will reach the stars. In fact, had Voyager 1 been pointed toward Proxima Centauri, the nearest star to us aside from our own sun, the journey would take around 76,000 years. If this sounds so long a timescale as to render interstellar travel impractical, here's the point: the universe is around 14 billion years old and some civilizations could have a head start of millions or even billions of years on us! This being the case, even if alien technology never exceeded the level of our own, the cosmos could be full of extraterrestrial probes similar to our own Pioneer or Voyager craft. This was the thrust of a scientific paper titled "On the Likelihood of Non-terrestrial Artifacts in the Solar System," by Jacob Haqq-Misra and Ravi Kumar Kopparapu, published in 2011 in the International Academy of Astronautics journal, *Acta Astronautica*. This isn't a one-off paper by maverick scientists. Volume 431, number 7004, of the prestigious science journal *Nature*, published on September 2, 2004, contained two features on the same theme, titled "Inscribed Matter as an Energy-Efficient Means of Communication with an Extraterrestrial Civilization," by C. Rose, and "Astrobiology: Message in a Bottle," by W. T. Sullivan. The latter began by stating: "Extraterrestrial civilizations may find it more efficient to communicate by sending material objects across interstellar distances rather than beams of electromagnetic radiation."

None of the material set out so far in this chapter proves that we've been visited by aliens, but it shows that interstellar travel is in no way incompatible with the laws of physics and requires only a level of technology that we already possess.

It's an odd part of the story that while the UFO community might see Burroughs and Penniston as standard-bearers for their theories about alien visitation, the two key witnesses won't play ball. Indeed, in one sense Penniston is as cynical about aliens as he is about police cars, lighthouses, or burning trucks of manure. In a mirror image of most UFO conspiracy theories, he talks about alien visitation not in terms of what was being covered up but in terms of it being the cover story:

> Well, it is as much as I would have guessed. With the cover story going out about UFOs and lights in the sky, alien spacecraft being fixed, and all the other rubbish. It was just another UFO story, to be disbelieved. So the debunkers and people, who were not in the know, went on about such nonsense. Simple as that it is. It is easy to see who was not in the know and who was, following the disinformation that was put out. Even people who have tried to write themselves into the story went with parts of the cover story. Amazing.

Burroughs, too, talks about aliens in terms of a cover story: "I believe the alien part was planted by both governments at the beginning, to cover up what was really going on, both by people who knew what they were doing and people who did not."

Pressed for a definitive view on the twin questions of the existence of extraterrestrial life and the possibility of alien visitation, Penniston says this: "I believe the odds are that the universe is brimming with life. I even believe there could be life with self-awareness in the universe. Do I believe that we are being visited by extraterrestrials? No, I do not see any evidence of that."

INTERACTION WITH CURRENTLY HIDDEN DIMENSIONS AND PARALLEL UNIVERSES

Parallel universes and hidden dimensions may be sci-fi staples, but is there any substance to such ideas? The answer, surprisingly, is that there are plenty of theoretical physicists who take such things seriously. Indeed, a particular branch of theoretical physics—string theory—actively *requires* that hidden dimensions be considered, in order to be viable.

We are used to perceiving the three dimensions (length, width, and depth) of space and the fourth of time. However, the mathematics of string theory suggests that in addition to the three spatial dimensions and the fourth dimension of time, there should be an additional six spatial dimensions. An extension of string theory called M-theory suggests that there are a total of eleven or maybe twenty-six dimensions. Intriguingly, in this theoretical model some degree of interaction between universes is permitted.

The likelihood is that we can no more perceive of such things than the inhabitants of the fictional two-dimensional world of Flatland could perceive a third dimension. In *Flatland,* written in 1884, the narrator (a square, who lives alongside other shapes on the two-dimensional Flatland) is visited by a three-dimensional entity, a sphere. He cannot conceive of the true nature of the sphere until he sees the three-dimensional world from which it has come (Spaceland) for himself.

Closely allied to the idea of hidden dimensions is the concept of a parallel universe, multiple universes, or even an infinite number of parallel universes—sometimes dubbed the multiverse. In such theories, our universe is just one of many universes: "one bubble floating in a sea of bubbles."

It's important to add that at the time of writing the existence of any extra dimensions—while an essential part of string theory—is unproven. But with theoretical physicists such as Michio Kaku speaking out on such matters and with scientists at the Large Hadron Collider (the world's largest and most powerful particle accelerator) looking seriously at such things, we should at least consider the possibilities here.

Much of the literature related to parallel universes and hidden dimensions is highly technical, speculative, or pseudoscientific. For that reason, we do not propose to delve further into this subject, but readers wanting a short primer from a reputable source should read the essay "Are There Other Dimensions?" on the Large Hadron Collider Web site. Indeed, it's the Large Hadron Collider that provides our best hope of discovering the existence of hidden dimensions. But until the existence of parallel universes and hidden dimensions is proven, theories about something from such realms interacting with our own world, e.g., in Rendlesham Forest, can only be speculation.

TIME TRAVEL

The final exotic theory considered in this chapter is time travel. As with parallel universes and hidden dimensions, what might at first sound like science fiction is taken rather more seriously by some theoretical physicists than many might suppose.

As we explained in the section on hidden dimensions, time is actually the fourth dimension. It is inextricably bound with the three (or eleven or twenty-six!) spatial dimensions, which is why physicists tend to use the term "space-time." Definitions of time vary, but a good way of thinking of it is that it's the "rate of change" of the universe. This reinforces the inextricable linkage, because time cannot exist without space and vice versa.

We are all time travelers. That is to say, all of us are moving forward in time. However, time doesn't flow at a constant rate; it's relative. There are a couple of reasons for this. First, gravity pulls on space, but because space and time are inextricably linked it also pulls on time. The closer a clock is to a source of gravitation, the slower it runs (gravitational time dilation). In addition, time passes more slowly the faster you travel (relative velocity time dilation). It was Albert Einstein who first theorized about this in his Theory of Relativity. Einstein's theories have since been verified by experiment, most famously involving comparisons of atomic clocks at different heights and speeds. Atomic clocks run faster at higher altitude (because there's less gravitational pull) and run slower at higher velocities. Experimentally the differences were measured in nanoseconds, but it's possible to envisage such effects becoming profoundly noticeable. The best example of this is the so-called twin paradox, where one twin takes a round-trip on a fast-moving spaceship. Such a twin would return younger than the twin who had remained on Earth.

If we all travel forward in time (albeit at different rates), can we travel back in time? First of all, it's worth making the point that we can see into the past. In fact, we do so all the time. The speed of light, in a vacuum, is around 186,000 miles per second (or 671 million miles per hour!). This makes no difference in living our day-to-day lives but is profoundly important and noticeable when it comes to astronomy. When you look at the moon, you are seeing it as it was around 1.3 seconds ago, because that's how long it takes light to travel from the moon to the Earth. The light from the sun takes a little over eight minutes to reach us. The light from the nearest star to us aside from our sun takes a little over four years to reach us. The deeper into space we probe, the further back in time we go. Some stars in our own galaxy are tens of thousands of light-years away, and if you look at Andromeda (a neighboring galaxy) you are seeing it as it existed over 2 million years ago! The Hubble Space Telescope is able to see objects so far away that it's effectively looking billions of years into the past, seeing the universe as it was only a short time after the big bang. Bizarre as it may sound at first, what this means is that many of the most distant stars visible to the Hubble Space Telescope no longer exist. Some red giants will have exploded and become supernovae; other variable stars will have become white dwarves. Hubble is a sort of time machine, showing us the universe not as it is but as it was.

Is any form of more practical time travel possible? There's a wealth of speculative material out there, but much of it is pseudoscience and/or New Age mumbo jumbo. However, among the noise there do seem to be a few signals worthy of further attention. Theoretical physicists have come up with a number of suggestions as to how time travel might be possible. These range from rotating black holes and wormholes through to cosmic strings and faster-than-light travel. We don't propose to delve too deep into any of these theories here, as it's beyond the scope of this book. The important point here is simply that while there's a lot of science fiction and pseudoscience about time travel, there are some theoretical physicists who give credence to the idea that it may be possible. Ronald L. Mallett is one such individual. Professor of physics at the University of Connecticut, Mallett is actively researching time travel. On his faculty Web site he includes "time travel" on his list of "primary research interests" alongside (slightly) more conventional topics such as black holes and quantum cosmology.

Some scientists look with skepticism not just at research such as Mallett's but at the field of theoretical physics more generally. After all, a central tenet of science is that theories must be tested and validated by experimentation. Moreover, they must be repeatable. Any theory that fails these tests must be discarded—though the validity of an experiment that appears to disprove a theory can itself be challenged. But much of theoretical physics is—well—theoretical. Saying something like, "We could test that by flying a spaceship at near light speed in the vicinity of a black hole," doesn't cut it with some scientists, because it simply can't be tested. Mallett, at least, is trying to address such concerns and his "Space-time Twisting by Light" project is, despite the technical-sounding name, a project aimed at constructing a time machine.

In a particularly fascinating twist, Dr. Mallett has speculated on the only practical way of sending a message back in time: using binary code; to use binary code through time using circular patterns of light to twist space-time, using light to manipulate time. The energy of the light beams will produce a gravitational field strong enough to drag a spinning neutron through time. Dr. Mallett says: "Let's say I'll call the spin up a one (1) and I'll call a spin down a zero (0). So imagine sending a stream of neutrons with spin up, spin up, spin down, spin up . . . what do you call that? It's binary code! So by using the spin of neutrons I can send the binary code which can be translated into a message."

Dr. Mallet believes that the first transmission of information to the past would be done via the transmission of binary code. It would be given to a receiver in the past. It's interesting that binary code has already cropped up in relation to the Pioneer 10 and Pioneer 11 spacecraft, as the consensus seems to be that this would be a logical way for either extraterrestrials or time travelers to communicate with us.

Can we come at the problem the other way around? In other words, instead of asking whether time travel is possible, can we look for evidence of it? There are, inevitably, a number of cases cited on the Internet purporting to show "out of place" individuals who might be time travelers from the future. Perhaps the best-known example is a clip from a 1928 film about the premiere of the Charlie Chaplin film *The Circus,* in which a woman appears to be using a cell phone. In another example, a man in what appears to be designer sunglasses and a modern T-shirt is visible in a 1941 photograph. But in all such cases, either the provenance of the photo or film is in doubt or the apparent anomaly is disputed. My personal favorite is the classified advertisement that appeared in a 1997 issue of *Backwoods Home Magazine*, which read: "Wanted: Somebody to go back in time with me. This is not a joke. P.O. Box 322, Oakview, CA 93022. You'll get paid after we get back. Must bring your own weapons. Safety not guaranteed. I have only done this once before."

Long since exposed as having been a joke "filler" written by one of the magazine's staffers, John Silveira, the advertisement became an Internet sensation and inspired—among other things—a 2012 movie comedy titled *Safety Not Guaranteed*.

The apparent absence of time travelers from the future visiting us leads to a temporal version of Fermi's "Where are they?" paradox. But as with the Fermi Paradox, there are several possible responses to the apparent absence of time travelers among us. These include the fact that time travel may be strictly controlled and that any time travelers among us will not interfere with events and will disguise their presence. This would be necessary to avoid such things as the "grandfather paradox," when someone goes back in time and kills their grandfather before he had married. But if they did this, they would never have been born—so how could they go back in time in the first place? Ironically, a possible solution to the grandfather paradox involves parallel universes, but this mixing of speculations is about as far as we can delve into this subject without getting bogged down in concepts that start in theoretical physics but soon end up as philosophical debates.

None of the speculative discussion in this chapter has delivered a definitive explanation for the Rendlesham Forest incident. But in relation to exotic theories about extraterrestrial visitation, interdimensional intrusion, and time travel, it was important to at least ask the question "Are such things possible?"

11. THE STORY GETS OUT

For all our modern obsession with open government and freedom of information, we sometimes forget that there are (usually) good reasons for secrecy. Examples of secrecy in the shadowy worlds of defense and intelligence might include information about nuclear weapon security or the identity of an informant in al-Qaeda. The potential consequences of such information falling into the hands of the wrong people are obvious and catastrophic.

In 1980, when the Rendlesham Forest incident took place, military secrecy was very different when compared to the situation today. While the US Freedom of Information Act was enacted in 1966 and came into effect on July 5, 1967, exemptions covering defense and national security were wider than they are today. So far as the United Kingdom was concerned, the matter was a complete non-issue. The United Kingdom's Freedom of Information Act was not enacted until 2000, and it was January 1, 2005, before the full provisions came into force.

Back in the eighties the MoD was an inherently secretive organization. The default position was to say nothing, and only a very few parts of the organization interacted with the public or the media. Legislation such as the Public Records Act did make provision for the release of information to the United Kingdom's National Archives, but the rule of thumb was that the most recent document in a file would have to be thirty years old before a file could be released. Even then, there was ample provision, with particularly sensitive issues, to extend this to fifty or one hundred years or indeed to ensure that files never saw the light of day at all.

Aside from legislation, the lack of several things that we take for granted today made it easier to keep secrets in 1980 than is the case now. The Internet (at least as we know it) didn't exist, and neither did social-networking sites such as Facebook and Twitter or whistle-blower sites like WikiLeaks. Moreover, the cell phone was yet to be the ubiquitous item that it is today. People were far more insular, and when the Rendlesham Forest incident occurred the chain of command must have felt that the events were unlikely to become public knowledge. They were wrong.

It's not possible to be certain exactly how soon knowledge of the Rendlesham Forest incident first leaked out, but the seeds seem to have been sown on New Year's Eve 1980, mere days after the event, by an indiscreet US airman who has been given the pseudonym "Steve Roberts" but whose identity is known to us. Roberts told a version of the story to a local musician called Chris Pennington. Pennington's partner, Brenda Butler, had a long-standing interest in UFOs and the paranormal and was soon on the case. Nor was "Steve Roberts" the only person talking. Several US airmen were discussing the incident with friends in the local community. Some of these people had been directly involved and others had not. Some accounts were relayed directly and others were overheard conversations, where only a few words and phrases were caught. To further add to the confusion, a number of local civilians had seen UFOs, too, and these people were talking. Much of this interaction was taking place socially, in a handful of local bars frequented by both off-duty military personnel and local civilians. Inevitably, tongues were at their loosest when alcohol was involved, and the net result of all this was that a decidedly garbled version—or, rather, versions—of the story began to leak out here and there, with a number of different people having different pieces of the puzzle. Just about the only things that were clear was that something highly unusual had occurred, that a UFO sighting was at the heart of it, and that numerous military personnel at the twin bases of Bentwaters and Woodbridge had been involved.

There were several independent sources for all this information. One of the most prolific was Larry Warren, a young airman based at Bentwaters/

Woodbridge, who was one of the first whistle-blowers on this case and was—in the early years—given the pseudonym of Art Wallace.

Larry Warren is one of the most controversial figures associated with the entire incident. Warren had only recently completed his training and arrived at the base only very shortly before the incident took place. He claims that on the same night as Halt's encounter he was ordered off his regular guarding duties and assigned to a makeshift group tasked with fueling the light-alls and taking them out into the forest. Warren then recalls that they dismounted from the vehicle, made their way through the forest, and emerged into the farmer's field. There he says that there was an illuminated patch of mist or fog that seemed to pulsate, as if there was an indistinct object of some sort present. Warren states that there were perhaps as many as forty men in the field, some American and some British. At this point he says that he heard a voice over the radio say, "Here it comes—here it comes," at which point a small red ball of light came in at high speed from the direction of the coast and hovered above the illuminated area of the field. The red ball of light then silently exploded with a flash that was so bright, it hurt his eyes. When he looked again, the ball of light and the illuminated area had disappeared and in its place was a structured craft, perhaps thirty feet across and twenty feet high. The object was roughly pyramid shaped, pearl white in color with some blue lights on the underside. It was translucent and Warren says that he saw indistinct figures inside, which appeared to be floating. Various personnel were taking photographs and filming the event, which appeared to have been expected. There were various other lights in the sky and beams being fired down at the ground.

Warren goes on to claim that at this point a senior officer, possibly Halt but more likely Williams, stepped forward and began to speak to the figures, perhaps normally or possibly telepathically—he wasn't close enough to be sure. Details of the alleged communication are sketchy, but over the years one claim that's surfaced is that the phrases "electronics division" and "a part from another world" were used, prompting UFO believers to come up with theories revolving around a damaged alien spacecraft being repaired by the USAF, presumably as part of some alien liaison program (which is actually quite a common belief in the UFO community).

Warren claims that some men had fled in panic by then and that others—himself included—were then ordered to leave the scene. He says that later on he was subjected to a post-incident debriefing during which witnesses were checked with a Geiger counter, shown films of UFOs, and told that there was an alien presence on Earth, of which the government was well aware. The assembled airmen were given watered-down statements of the event to sign and the two briefers implied there would be adverse consequences if any of the witnesses discussed what they'd seen. Warren apparently asked what would happen. "Bullets are cheap," one man replied. "Yeah, they're a dime a dozen," his colleague chipped in.

In later years, Warren was to claim that he had been snatched by sinister "Men in Black" type individuals, taken to a massive underground facility below the Bentwaters facility, and interrogated.

What are we to make of all this? In some respects, Warren's story contains elements that we have encountered in the accounts of other witnesses. But many parts of the story, including the UFO in the field, communication with aliens, and the underground facility, are unique to Warren's account. In evidential terms, Warren's story is problematic for several reasons. Most important, none of the other key witnesses recall seeing him at any stage during the various encounters. Conversely, just about all the other witnesses are able to point to one, two, or more people who were with them at the time, so that there's corroboration: Burroughs encountered the UFO with Penniston; Halt encountered the UFO with Englund, et cetera. Another problem is that there's no witness statement from Warren and neither do any of the other witness statements mention him. In addition, Warren's story has changed many times over the years. These points alone would not be showstoppers (we've seen that several key players wrote no witness statements, and there have certainly been some discrepancies in other witnesses' accounts) and one would expect recollections and opinions to change and evolve over thirty years, but Warren's story has changed more than most. The key problem, though, is the first one. Warren's story is essentially a "stand-alone," whereas the accounts of everyone else mesh together to create a bigger picture.

At first, Halt and others suggested that Warren had not even been posted to Bentwaters/Woodbridge at the time of the incident, but this was not correct and documentary evidence shows that Warren had arrived on or around December 11, 1980. However, even by Warren's own admission, parts of his story were apparently not his but the experiences of another witness—Sergeant Adrian Bustinza—who did not wish to go public at the time but who enlisted Warren's help to get his story out.

It's perfectly possible that Warren is entirely sincere in his recollections. He has, by his own admission, undergone regression hypnosis in an attempt to probe his memories. But there's strong evidence to suggest that far from helping to recover lost memories, this technique can implant false ones. And bear in mind Halt's statement that "drugs such as sodium pentothal, often called a 'truth serum' when used with some form of brainwashing or hypnosis, were administered during these interrogations, and the whole thing has had damaging, and lasting, effects on the men involved." Now, Halt was discussing "five young airmen" whom he doesn't name but who are clearly the five whose statements are reproduced in chapter 7, i.e., Buran, Penniston, Burroughs, Chandler, and Cabansag. However, if some witnesses were interrogated in this way, can we really say that Warren's story is that outrageous? It would be easy to fall back on the "you can't prove a negative" defense, though Sagan's "extraordinary claims require extraordinary evidence" view is arguably applicable here. Alternatively, we can fall back on the assessment used by intelligence analysts the world over, when confronted with an intriguing but uncorroborated piece of intelligence: "interesting, if true."

Whatever the scope of Warren's involvement, most of the witnesses acknowledge that he was one of the initial whistle-blowers—maybe even the first one—and that he was therefore responsible for getting the story out before the other witnesses could do so, due to their active-duty status in the USAF. Most of the other key witnesses could only discuss the events openly after their retirement in the early 1990s, whereas by 1983 Warren had left the USAF.

We previously mentioned another whistle-blower, Steve Roberts (another pseudonym), who was actually the first to go public in and around the pubs in the area. According to Penniston, Burroughs, and other key witnesses, Steve Roberts was actually a sergeant named J. D. Engels, who worked in the 81st Security Police Squadron, Reports and Analysis Section, within the Security Police Administration branch. The section that he worked in had access to all Law Enforcement and Security Police blotters, statements, and incident and complaint reports. He also knew Penniston (who assumed duties as NCOIC, 81st Security Police Plans and Programs, in February 1981) quite well, as his office was only a hundred feet from the Reports and Analysis office. Penniston recalls that Engels was always trying to pump him for information and openly admitted he had read the blotters and incident reports right after they were pulled from the Central Security Control and the LE desk.

An additional whistle-blower was David Potts (yet another pseudonym). David Potts was a UK radar operator based at RAF Watton, who spoke to several UFO researchers and stated that an "uncorrelated target" had been tracked over Bentwaters/Woodbridge at some time in the last week of December 1980. He went on to say that shortly afterwards a group of US military officials visited the radar base, confirmed some basic details of the UFO encounter, and asked to see the radar reports. There is a strong suggestion that some radar tapes and/or documentation was then removed, and readers will recall from chapter 8 that the MoD and RAF documents state that the cameras and films were faulty on the night(s) in question and that no data exists.

Into all of this confusion stepped a small but dedicated group of individuals who would describe themselves either as ufologists or as paranormal investigators. Such people are often regarded as eccentrics, and while the UFO community certainly has more than its fair share of crackpots, charlatans, and cultists, there are some highly intelligent, diligent, and levelheaded characters involved, too. A number of such people—Paul Begg, Bob Easton, Harry Harris, Ian Mrzyglod, Norman Oliver, Jenny Randles, Mike Sacks, Peter Warrington, and others—played a part in all this, but it was Brenda Butler and her research partner Dot Street who were to do most of the legwork in trying to piece together what happened. Many of these people were either members of or closely involved with, the British UFO Research Association (BUFORA). Formed in 1962, BUFORA describes itself as one of the oldest and most consistently active UFO research and investigation groups in the world. In October 1981 BUFORA discussed the case at one of their regular meetings. As it transpired, the most important thing about the meeting wasn't the discussion itself but the fact that Dot Street met Jenny Randles, who had recently been appointed as BUFORA's Director of Investigations. From this point onwards, Jenny Randles would work more closely with Brenda Butler and Dot Street on piecing together what had happened.

For all of this, there were few tangible results. Jenny Randles pulled together the various disparate pieces of information and wrote up an account of the case that was circulated to the forty or so readers of a UFO newsletter. This article was subsequently published in the magazine *Flying Saucer Review*, in 1982, where it reached a slightly larger audience. Another article (summarizing the collaborative researches of a paranormal research group called Probe and a group called the Swindon Center for UFO Research and Investigation) appeared in a small publication called *PROBE Report*. But this was small-scale stuff—these sorts of newsletters and magazines only reached a very specialist audience and weren't available on newsstands. A more significant piece of publicity came in March 1983 when *Omni* (a now-defunct magazine that carried features and news items on science, science fiction, and the paranormal) printed an account of the story. But all of these stories were a mixture of fact and fiction, and while the *Omni* story contained a quote about the incident from the Base Commander, Colonel Ted Conrad, there were no documents that would support the story. Everything rested on anonymous (or pseudonymous) hearsay.

The persistence of Brenda Butler and Dot Street was beginning to cause some concern in the MoD. On October 25, 1982, Squadron Leader Donald Moreland (who had spoken to Charles Halt in the immediate aftermath of the 1980 sightings and who advised Halt to write up a report for the MoD) wrote to the MoD's UFO project as follows:

1. Under cover of reference A I forwarded you a copy of the Deputy Base Commander's report concerning some unexplained lights and sighting on 27/29 December 1980. Some time after the incident I was approached by two women who claimed to be UFO investigators, but I refused to confirm or deny their claims. A week ago I was telephoned from New York by a Mr Eric Mishara from Omni Magazine. He claimed he was serious UFO investigator and wanted to write an objective article about the incident. I told him that whoever wrote the article he described to me must have had a vivid imagination.

2. I have now managed to obtain a copy of the article and enclose a

copy for your information. The magazine is called "The Unexplained"
published weekly by:

ORBIS Publishing Ltd
Orbis House
20/22 Bedfordbury
London WC2N 4BT

The article was in Volume 9 Issue No 106.
　3. I now anticipate a flood of enquiries and would be grateful for
some guidance on MOD policy concerning UFO's.

The article in *The Unexplained* had been written by Jenny Randles.

Peter Watkins at the MoD responded to Moreland on November 9, 1982. Watkins started out by providing some general "lines to take" on the MoD's UFO policy, as Moreland had requested. He went on to provide a specific line to take on Rendlesham, stating that they were aware of a report but that it had not been judged to be of "defence interest." But what clearly alarmed Peter Watkins most was the fact that the article in *The Unexplained* had mentioned the possibility that the events had been a cover story for the crash of an aircraft carrying nuclear material. Watkins referred Moreland to previous denials that had been given in the UK Parliament, both in relation to the general issue of nuclear weapons accidents on UK soil and in relation to the specific case of the 1956 accident at Lakenheath, mentioned in chapter 9.

Jenny Randles, too, was pressuring the MoD. In one letter, dated February 28, 1983, she set out her own imaginative view of the Rendlesham Forest incident: "I would add that the story behind these events indicates that there was contact between military sources and another intelligence (which is not alien spaceships in the nuts and bolts sense) but which is an indigenous intelligence to planet earth which in fact is way beyond us in terms of most capacities and therefore represent the real rulers of our world." But she had also asked for a copy of any report on the incident held on file, and while the United Kingdom did not yet have a Freedom of Information Act and government files on any subject were normally closed for thirty years, MoD officials correctly surmised that pressure was growing. Pam

Titchmarsh from DS8 (the section in which the UFO project was embedded at the time) wrote to the Senior RAF Liaison Officer (SRAFLO) at HQ Third Air Force, RAF Mildenhall. In her letter, dated May 13, 1983, she wrote:

> You will see that she [Jenny Randles] has now written again seeking further information about the incident and in particular has requested a copy of the report held on our files. The only report we have is that prepared by Lt Col Halt the Deputy Base Commander at RAF Woodbridge and I am therefore writing to ask you to seek the views of USAF to the disclosure of that report or a sanitized version of it. If the USAF would only be prepared to allow release of a sanitized version it would be helpful to know which parts they would wish me to delete. In addition, I would be grateful to know whether the USAF would be willing for me to say that they did investigate the incident.

It's easy to see how conspiracy theorists would interpret talk of sanitized versions of reports and deletion of text, especially given what happened in the case of the five USAF witness statements from Buran, Chandler, Penniston, Burroughs, and Cabansag.

Pam Titchmarsh and the SRAFLO discussed this on the telephone on May 17, but the precise details of this conversation are unknown. No note of this conversation was kept, though it is referred to in a May 18 letter from the SRAFLO, Wing Commander J. R. Davies, to Pam Titchmarsh:

> 1. Thank you for your letter at Reference A and enclosures. I said in the telephone conversation at Reference B that it will be some little time before we can get a decision on the release of the report by Lt Col Halt. In fact, the decision to allow the release may have to come from Secretary of State for Defense's office, particularly if any security or intelligence implications are read into the reported sighting.
> 2. I will let you know of developments as they occur.

We cannot overstate the importance of this brief statement. It's a telling indication of what was going on behind the scenes, that exchanges of letters between middle-ranking civil servants and military officers suddenly warranted an escalation to the Secretary of State for Defence!

In fact, developments were occurring rapidly and the storm was about to break.

For all the efforts of British ufologists, the key breakthrough came as a result of American researchers. Several of the people who played a key part in bringing the Rendlesham Forest incident out of the shadows were US researchers Ray Boeche, Scott Colborn, Lucius Farish, Larry Fawcett, Barry Greenwood, and Robert Todd. Fawcett, Greenwood, and Todd were all members of an organization called Citizens Against UFO Secrecy (CAUS). US researchers had an advantage over their UK opposite numbers when it came to their research and investigations—and, specifically, their interaction with the government and the military. It was the Freedom of Information Act.

On the basis of the bits and pieces of information that were already in the public domain, CAUS had made a number of carefully worded Freedom of Information Act requests relating to the case. Unlike asking open questions (which they'd also tried but which—as with the UK researchers' experience—tended to meet with blanket denials), such requests also covered documents. There was not much of a paper trail, but what did exist, of course, was Charles Halt's one-page report to the MoD.

Halt, as we saw in chapter 8, had been ordered to liaise with RAF Commander Donald Moreland and send a report of the events to the UK MoD. Though worried that such a report would effectively end his career, Halt was beginning to think that he had heard the last of the matter. The probing of people such as Butler, Randles, and Street had been a nuisance, but Halt had dealt with this in a polite but effective manner, giving them little to go on. The CAUS FOI requests were to change all this.

The initial CAUS FOI requests failed. A response sent to them on April 23, 1983, by Colonel Henry J. Cochran at Bentwaters read as follows: "Reference your letter dated April 14, 1983, requesting information about unknown aircraft activity near RAF Bentwaters. There was allegedly some strange activity near RAF Bentwaters at the approximate time in question but not on land under U.S. Air Force jurisdiction and, therefore, no official investigation was conducted by the 81st Tactical Fighter Wing. Thus, the records you request do not exist." This response was inaccurate, in that an official investigation *had* taken place (two, if one counts the USAF and MoD investigations as separate) and records *did* exist.

CAUS tried again on May 7 but this time sent their request to the Headquarters of the 513th Combat Support Group—a unit that provided document management support services to 3rd Air Force, RAF Mildenhall.

Halt recalls receiving a call from Colonel Peter Bent at HQ 513th CSG, informing him that as a result of the FOI request they had located a copy of his memo and were going to have to release it. Bent (who was a friend of Halt's) was forewarning Halt, out of both friendship and professional courtesy.

Ironically, the US authorities had initially been unable to locate a copy of Halt's memo in their own files and it was Wing Commander Davies, the SRAFLO at HQ 3rd Air Force, who provided the copy, following his earlier discussions with Pam Titchmarsh at the MoD. Halt recalls asking Bent to "burn it," saying that his life would never be the same again. Bent didn't burn it, but Halt's prediction turned out to be correct.

On June 14, 1983, Bent responded to the FOI request in a letter to Robert Todd of CAUS. The letter contained the following statement: "It might interest you to know that the U.S. Air Force had no longer retained a copy of the January 13, 1981, letter written by Lt. Col. Charles I. Halt. The Air Force file copy had been properly disposed of in accordance with Air Force Regulations. Fortunately—through diligent inquiry and the gracious consent of Her Majesty's government, the British Ministry of Defense, and the Royal Air Force, the U.S. Air Force has provided a copy for you."

The cat was out of the bag. There had certainly been a rather fortunate (or unfortunate, for some) and convoluted chain of circumstances: US service personnel had leaked information to British UFO researchers. This gave American UFO researchers enough of a lead to submit a sufficiently well-targeted FOI request. The US authorities couldn't locate a copy of Halt's memo but acquired one from the British.

Now the UFO researchers had a copy of Halt's memo, it was much easier to interest the mainstream media in the story, as it had suddenly been elevated from hearsay to an officially confirmed account. It was only a matter of time before the story broke.

The CAUS team that had acquired the Halt memo thought it only fair that they pass a copy to their UK counterparts, on the basis that it was their initial legwork that led to the acquisition of sufficient information for CAUS to craft their FOI request. Despite the rivalries that exist within the

UFO community, there can be a certain camaraderie, too, and the CAUS team was determined to do the right thing. However, once the Halt memo had been acquired, there was no real sense of where next to take the investigation. The CAUS team and the various UK researchers had no real strategic vision or plan, and inevitably this led to people going their own way.

Burroughs and Penniston have mixed views about ufologists. While admiring the dedication and tenacity of some of the individual researchers who worked so hard to uncover documents on the incident, they are more critical of the UFO community as a whole. Penniston puts it this way, in describing their conclusion-led wishful thinking:

> I believe that in general, the UFO community as a whole wants answers. I also believe there are a significant portion of them who deal in wishful thinking. They are the ones who can take a picture of a cloud and call it an UFO, or the ones who see lights in the sky and the first words out of their mouths are that it is a UFO (ET/Alien), when in actuality an informed observer could easily identify it. Their UFO is an observation of one of the following type of possibilities: a manmade object, a star/planetary body, or other natural occurring phenomena— all completely identifiable by a trained observer. But their need to not be alone [in the universe] seems to overwhelm their own judgment.

At this point, several members of the UK team (or loose alliance, as it might be more accurate to say) sold the story to the press. This irrevocably split the US/UK "alliance" of UFO researchers and indeed caused controversy and bitterness among the UK researchers (and the wider UFO community) that linger to this day. There are debates and disputes about whether one of the UK researchers contacted the media or the media had become aware of the story and reached out to the UFO community, disputes over how much was paid, disputes over which researchers received payment (and how the money was divided) and which didn't, and even a dispute over whether the Halt memo was sold or the *story* was sold and the Halt memo was freely given as part of this story. While the facts are known to us, they're not really relevant here. What really matters is who got the story. The paper concerned was the *News of the World*.

The *News of The World* was a popular UK tabloid that came out on

Sundays. It was owned by media mogul Rupert Murdoch and famously closed in 2011, following a notorious phone-hacking scandal. In 1983 it was the United Kingdom's bestselling Sunday paper. It certainly had its critics—it was nicknamed News of the Screws and Screws of the World, due to the number of sex scandal stories it published. That said, it actually broke some major stories, exposed corruption, fronted several high-profile public-interest campaigns (e.g., a 2000 anti-pedophile campaign that led to the public gaining access to data on the sex offenders' register [Sarah's Law]), and won the British Press Awards' prestigious Newspaper of the Year award in 2005.

The journalist who broke the story was veteran reporter Keith Beabey. Beabey had spoken to various members of the UK team but was desperate to speak to witnesses, not ufologists. Like the good journalist he is, he hounded Halt resolutely,

The story was published on October 2, 1983. The *News of the World* editor decided to make the story the main story on the front page—one of perhaps only two or three occasions on which a UK national newspaper had made a UFO story their main, front-page headline. "UFO Lands in Suffolk," screamed the main headline. "And that's OFFICIAL" ran a smaller headline, underneath. At the top of the page were three newsy bullet points that read "Colonel's top secret report tells the facts," "Mystery craft in exploding wall of color," and "Animals flee from strange glowing object." The best way to characterize the story is to say that parts of it were true and parts were not. The parts based on the Halt memo were generally pretty factual, but by focusing on some of the more sensationalist claims the account veered from fact into fiction.

The chain of command seemed totally unsighted. There's no evidence to suggest that senior USAF commanders knew the story was about to go public, let alone took any action to forewarn any of those involved what was about to happen. The witnesses themselves certainly had no advance warning of any of this and were far from pleased. Burroughs describes his feelings thus: "I was not happy that the story got out. It was something I did not understand and was not prepared to deal with." In the media, there are few prizes for second place. Competition over major stories can be brutal, and when one paper "scoops" everyone else by breaking a big story those who lost out on the story either quickly jump on the bandwagon or trash the story. The paper that first gets the scoop can fight back, but if it's a Sunday paper this is much less likely, because the daily newspapers have six whole days in which to run stories, while a Sunday paper such as the *News of the World* comes out just once a week. So if the media trash a Sunday paper's story in the early part of the week following publication, they won't have a chance to respond for several days, by which time the whole story will be looking rather dated: "Today's newspaper is tomorrow's fish-and-chip paper," as the old UK saying goes.

And that's precisely what happened. A couple of Sunday papers actually got sufficient warning of the *News of the World* story to stop the presses and slip in their own account of events on the same day. But for the most part, it was the national daily papers that picked up the ball and ran with it. In their attempts to get a new angle on this, journalists contacted the Forestry Commission and sent reporters to the area to knock on doors and stop people in the streets. And that's how the media found Vince Thurkettle. As we saw in chapter 9, local forester Vince Thurkettle was the original proponent of the theory that the UFO sighting resulted from misidentification of the Orfordness lighthouse. Although his current position is that the lighthouse might have played a part in events but couldn't possibly explain events that lasted several hours and spanned two nights, the lighthouse theory was seized upon by some of the media who had lost out to the *News of the World* on breaking the story. Much of the "anti-UFO, pro-lighthouse" media coverage that was published the week after the *News of the World* story may be attributable to "sour grapes" on the part of newspapers that were scooped, but could there have been something else at play?

One clue is the fact that the MoD was so tight-lipped. While the Defence Press Office received numerous queries following publication of the original story, they at first denied even having a copy of Halt's report when, in fact, it was the MoD that had provided the copy sent to the newspaper! While this may simply have been a case of the left hand not knowing what the right hand was doing (a more common problem in government than the public might suppose), it had the effect of preventing any awkward questions being asked and drove the media to focus on the only other place they had to go: a local man whose speculation about the lighthouse became—in some people's minds—a certainty. Was the MoD's evasiveness deliberate or an attempt to direct the media in a particular direction?

It's worth looking in detail at the MoD response. On October 3, the Monday after the story broke, Pam Titchmarsh from the MoD's UFO project wrote an assessment of the *News of the World* story. The relevant parts read as follows:

> The *News of the World* story appears to be one fabrication after another. Lt. Col. Halt has not spoken to anyone from the *News of the World* . . . the alleged interview with Sqn. Ldr. Donald Moreland is also a fabrication. He stated that "to the best of my knowledge Lt. Col. Halt is a very genuine person" but gave no details of any conversation he had had with Halt nor did he say "whatever it was, it was able to perform feats in the air which no known aircraft is capable of doing. . . ." The unfortunate point about the article is that MoD refused to comment on the grounds that it was a matter for the USAF while USAF were saying it was a matter for MoD.

It wasn't until October 6 that the MoD's UFO project sent the Defence Press Office any formal, written guidance. This consisted of a brief statement and some "defensive lines to take," i.e., responses to specific questions, which could be used if those questions were ever asked but which were not to be volunteered. The statement read as follows:

> I can confirm that the Ministry of Defence did receive a report from base personnel of a UFO sighting near RAF Woodbridge on 27 December 1980 (this was the report published by the *News of the World* on 2 October 1983). The report was dealt with in accordance with normal procedures, i.e., it was passed to staff concerned with air defence matters who examine such reports to satisfy themselves that there are no defence implications. In this instance MoD was satisfied that there was nothing of defence interest in the alleged sightings. There was no question of any contact with "alien beings."

This is either extraordinarily inept or a clever sleight of hand. The very first statement is wrong and repeats the incorrect date given in Halt's memo. The "staff concerned with air defence matters" is a reference to RAF radar experts who are routinely consulted on all UFO investigations, but the "no defence implications" assessment is weak to say the least. There's no mention at all of the DIS and their assessment that the radiation readings were "significantly higher than the average background." This is because the involvement of the DIS in UFO research and investigations had not yet been publicly acknowledged, but it also allowed the MoD to ignore addressing the radiation issue altogether.

The defensive question and answer material was even more illuminating:

Q1. *Did the US authorities investigate the incident?*
A1. *No. Once the report had been sent to the Ministry of Defence the US authorities carried out no further investigations. [Investigations of UFO reports in the United Kingdom are carried out by the Ministry of Defence; the USAF has no responsibility in such matters.]*

The first statement ("No") is demonstrably false, as evidenced by the various post-incident debriefs and—even if the precise circumstances of these are unclear—the five witness statements given by Buran, Chandler, Penniston, Burroughs, and Cabansag. The second part is possibly true but disingenuous, as around three weeks had elapsed before the US authorities formally notified the MoD.

Q2. *Was Col Halt told to stay quiet?*
A2. *No. Lt Col Halt has not been told to keep quiet about the incident nor has he been informed that his career could be in jeopardy.*

This seems to be true. While Halt was not eager to discuss any of this with the media or the public (and indeed went to some lengths to try to prevent the story getting out), there's no evidence that he was told to keep quiet. Similarly, while Halt believed that his career would be in jeopardy should the story be made public, there's no evidence to suggest he was ever told this. It was simply his personal view, which turned out not to be correct.

Q3. Was the object tracked on radar?
A3. No. No unidentified object was seen on any radar recordings during the period in question.

This was not correct.

It's not clear if the various false, partially false, or misleading statements were made because the people compiling the material weren't fully briefed, whether it was the consequence of a deliberate strategy to put the media off the scent, or whether it was due to some combination of these two factors.

In the final analysis, the long-running game of cat and mouse between the UFO community and the government had reached its climax. But who had won? The UFO community undoubtedly thought they had. After all, in the face of repeated denial and obfuscation ufologists had not only obtained the Halt memo but also gotten one of the United Kingdom's best-selling national newspapers to run the story on the front page. It seemed to many as if a famous victory had been achieved and that they had "gotten one over" on the people who—as many more conspiratorially minded ufologists believed—knew all about UFOs and were actively covering it up. On the surface, it did indeed seem as if the UFO community had won this round.

The reality is somewhat different. There's no doubt that the US and the UK governments tried to stop this story from coming out. It's self-evident that they made it extremely difficult for the UFO community. It's not clear if lies were deliberately told, because denials that the incident had taken place or that a paper trail existed may have been given in good faith, by people not briefed on these events. But at the very least, they had dissembled. However, once the Halt memo had been released to CAUS, the British and American authorities must have realized that it was only a matter of time before the story hit the mainstream press. After all, some aspects of the story had already been published in small, specialist UFO and paranormal magazines and newsletters. Accordingly, they must have been delighted that it was a tabloid that broke the story, as opposed to a broadsheet. They must also have been pleased that the story gave weight to some of the more sensational (and false) claims, at the expense of probing parts of the story where the authorities were vulnerable—particularly in relation to the radiation levels recorded at the landing site. Finally, the authorities must have been exceptionally pleased that—perhaps as a result of the bitterness that comes from having lost out on a major story—the rest of the media gave such prominence to skeptical theories about the events.

The MoD would have regarded this outcome as the best they could have expected, given the circumstances. The problem for the UFO community was that they have a very limited understanding of the media and tend to regard any publicity as good publicity. But while they got the story out, it was "spun" in such a way as to be at best sensationalist and at worst ridiculous. And within twenty-four hours the secondary media coverage was such that the agenda had subtly shifted. There were certainly some obvious and potentially awkward questions to ask: "How do the radiation levels recorded at the landing site compare to background level?," "What radar data exist for the nights in question?," and, "What other official documents exist?" Worse still, the media might have discovered that Halt's memo gave the wrong dates, making the US military look either incompetent or disingenuous. Critically, the media might have highlighted the fact that the senior US officer in Europe, General Gabriel, had removed evidence back to his HQ in Europe without even informing the UK authorities that this evidence even existed. Similarly, again highlighting something that could have driven a wedge between the United States and the United Kingdom, they might have discovered that the DIS had assessed the radiation levels as being "significantly higher than background" but had failed to pass this all-important assessment back to the US authorities, in a situation where several personnel had spent considerable time at the landing site. But none of this happened. The only questions that lingered were "Did all these highly trained US military personnel manage to misidentify a lighthouse?" and the (unstated) follow-up: "Are these guys nuts?"—for the government, it was indeed victory from the jaws of defeat. The question that then arises is this: were the authorities simply lucky? Did a series of events unfold that resulted in this potential ticking time bomb becoming a damp squib? Or was it the result of a carefully thought out media-handling plan?

12. RENDLESHAM RUMORS

In the last chapter we saw how the release of Charles Halt's memo and the subsequent publication of the story (or a version of it) ended nearly three years of secrecy and rumor. At first, it seemed as if the UFO community had won a famous victory. But perhaps by accident or perhaps by design, the truth was very different. The story that had actually been published was a mixture of fact and fiction. Some of the witnesses who had spoken out were the real deal, while others were wannabes. This latter category had undeniably been based at Bentwaters/Woodbridge but either had been very peripherally involved and had embellished their accounts or had not been involved at all but had picked up the story while on-base and attempted to write themselves into it. With some of these people the motivation might have been financial. Journalists do pay for stories and there's money to be made out of books, TV appearances, and lectures—though it takes a combination of luck and good business sense. With others, ego may have been the reason. But more sinister possibilities abound. Given the statements made about the use of threats, drugs, and hypnosis in some of the post-incident debriefings, could some of the accounts be a consequence of witnesses having been "messed with"—as some people in the UFO community put it? Or could one or more of the witnesses be a "plant"? In either of these two latter possibilities, the aim would be to muddy the waters, hoping that the bogus stories would detract attention from the real ones or be seen as so fanciful that the entire story would be ridiculed and discredited.

This was the downside of the publicity. If the situation before the *News of the World* broke the story was confusing, afterwards it was a nightmare. The UFO community had a number of aspirations, some more realistic than others. Some hoped that this would mark some sort of "tipping point," that the front-page story meant that "serious ufology" had attained a critical mass and that what in later years became known as "Disclosure" would come to pass. "Disclosure" is what one might call the ultimate fantasy, so far as ufologists are concerned: The President of the United States goes on television and announces an alien presence and apologizes for the government cover-up and Earth joins some sort of *Star Trek*–style galactic federation. Everyone lives happily ever after, especially ufologists, whom people look at admiringly, as they realize the person they previously thought was nuts had been right all along. Fantasies aside, the UFO community at least hoped that the publication of the story would lead to the UFO phenomenon being taken more seriously. This might have happened had the story that appeared concentrated on the radiation readings at the landing site, which was probably the strongest aspect of the story and was certainly the part of the story that was best evidenced by the paperwork available to the *News of the World*—essentially just the Halt memo at this stage, as the DIS assessment of the radiation levels had yet to emerge. But by leading with the story of contact with alien beings the paper had overplayed its hand. At the very least, ufologists hoped that the story would have the effect of encouraging other witnesses to come forward. This certainly happened, over the years, but once the story was in the public domain it also became a magnet for fantasists trying to write themselves into the story, either as witnesses or as experts. Getting the story into the mainstream media was indeed a two-edged sword.

With the preceding health warning in mind, in this chapter we're going to look at some of the most interesting rumors that emerged over the years in relation to the Rendlesham Forest incident. These are not united by any particular theme, so this chapter is something of a miscellany. But some of these stories have become inextricably linked with the Rendlesham story, so it's important to examine them carefully to see whether or not they're true.

JETS SCRAMBLED TO INTERCEPT RENDLESHAM UFO

One of the most intriguing rumors in relation to the case was that jets were scrambled in an attempt to intercept the UFO seen at Rendlesham Forest. The story involves a former RAF radar operator called Malcolm (sometimes abbreviated to Mal) Scurrah. In 1980 he was stationed at RAF Neatishead when an uncorrelated target was detected at a height of around five thousand feet. The target was not carrying an IFF (Identification Friend or Foe) radio transponder and was performing extraordinary maneuvers at speeds that seemed to exceed the capability of any known aircraft. A Phantom fighter/interceptor aircraft was vectored to investigate and the pilot reported that he saw a bright light when he closed to a distance of around half a mile. But the unidentified craft began to climb, and while the RAF Phantom pilot attempted to follow, the object's rate of climb was extraordinary and when it hit a height of over ninety thousand feet (significantly beyond the maximum ceiling of a Phantom) the pilot broke off his pursuit. At the time, the only aircraft capable of flying at such heights, so far as RAF personnel were aware, was the SR-71 Blackbird.

The following day some senior RAF officers visited and debriefed those who had been involved in tracking the mystery craft. Radar tapes were confiscated in a way that normally only happens in the event of an aircraft accident or a near miss between two aircraft.

While this incident occurred in 1980, it is not, in fact, linked with the Rendlesham Forest incident. Scurrah is not certain of the date but believes it was October or November. He emphatically does not believe it was as late as December and indeed was irritated when some researchers tried to shoehorn his sighting into the Rendlesham story. There are clearly similarities between Scurrah's story and the way in which it's alleged that radar tapes were confiscated in the aftermath of the Rendlesham Forest incident, but that's where the similarity ends. There was no flying on either night of the Rendlesham Forest incident and none of the witnesses saw or heard any aircraft or subsequently heard about any attempt to intercept the UFO with military aircraft. While this incident is fascinating in its own right, there does not seem to be any direct connection with the Rendlesham Forest incident.

POST-INCIDENT POSTINGS

It has sometimes been claimed that in the aftermath of the incident one method used by the chain of command to cover up the incident was to separate the witnesses. After all, there's strength in numbers and if those involved felt that they had a story to tell but were being let down by the chain of command they would perhaps be more likely to act if they were collocated. It's been claimed that after the incident those witnesses most closely involved with events were deliberately separated and posted to various different USAF establishments all around the world, making it much more difficult (in those pre-Internet and pre-Facebook days) for them to stay in touch.

This is a tricky one, because USAF posting policy can be Byzantine, and without knowing the exact tour lengths of all the witnesses and doing a comparative analysis with base personnel who were not involved it's difficult to be certain on this point. Needless to say, people are posted in and posted out of military establishments all the time, either on level transfer or on promotion. At the time of the Rendlesham Forest incident, some witnesses will have been at the beginning of their tour, some in the middle, and some at the end. In the months afterwards, a steady stream of those at the end of their tours will have left, but whether the flow was faster than usual, with people being posted who would normally have been expected to remain for longer, we can't be sure. Some of the witnesses certainly have the *perception* of there being a post-incident exodus. Staff Sergeant John Coffey, the Senior Security Controller on the first night, working in Central Security Control, though skeptical about some elements of the Rendlesham Forest incident (and therefore somebody with no dog in the race), has this to say: "My Blotter was pulled and classified SECRET by the Base Commander and almost all personnel that were involved in that incident and were PCSing that year had their PCS's moved up. I was supposed to PCS in June and it was pushed up to March, no reason given." "PCS" is the abbreviation for "Permanent Change of Station."

Penniston certainly agrees there was something decidedly odd here:

As far as accelerated time lines with assignments, I do think that was unusually high over the next six to eight months. Key witnesses were all rotated, with the exception of Colonel Halt and I—we both left the summer of 1984. I find that stranger than who left the base, we two alone remained. Higher than the normal, not including discharges for reason of inadaptability to the USAF, or criminal activity. Then of course some were just due to rotate with another assignment.

There is, therefore, some evidence to suggest that there were some unusual postings in the aftermath of the incident, though it's not possible to prove a causal link.

POST-INCIDENT SUICIDES

One of the most disturbing rumors in relation to the incident is that a young security policeman nicknamed Alabama (apparently on the basis that this was his home state) committed suicide after the events, because he was unable to handle what had happened. The story about the possible suicide was brought to the attention of Lord Hill-Norton, the former Chief of the Defence Staff who had taken an interest in the Rendlesham Forest incident and indeed in the UFO phenomenon more generally. He decided to probe further by asking a formal, written question in the House of Lords. A PQ can be asked in either the House of Lords or the House of Commons, depending upon whether the person asking the question is a Peer of the Realm or a Member of Parliament. In either event, PQs are taken extremely seriously and before word processors were widely available PQs used to be hand-delivered within MoD Main Building in menacing-looking green folders with red tags and were stamped "Ministerial business—to be given priority at all times." It was a "drop everything else" scenario and everyone in the chain took this seriously, from the desk officer researching the answer, to the head of division who signed off the answer, to the MoD minister (either the Secretary of State for Defence or one of the undersecretaries of state) who would actually give the answer. If an answer turned out to be wrong, the Speaker could summon the minister to Parliament, where a formal apology had to be delivered. Despite the cynicism of conspiracy theorists who believe governments routinely lie, if a minister was ever found to have deliberately misled Parliament it would be a resignation issue. Here is the formal record of the question and the answer concerning the alleged Rendlesham-related suicide, printed in the *Hansard* dated October 28, 1997:

Lord Hill-Norton asked Her Majesty's Government:

What information they have on the suicide of the United States security policeman from the 81st Security Police Squadron who took his life at RAF Bentwaters in January 1981, and whether they will detail the involvement of the British police, Coroner's Office, and any other authorities concerned.

Lord Gilbert MoD has no information concerning the alleged suicide. Investigations into such occurrences are carried out by the US Forces.

While saying that the MoD has no information about the suicide is not a denial that it occurred, the response is fair up to a point. However, while Bentwaters and Woodbridge were indeed USAF bases, they were on British soil and if there was a suicide at such a base the MoD certainly should have been aware and concerned about it. Again, we come back to complex questions of jurisdiction and primacy. As a further complication, police guidance on sudden deaths at military establishments in the United Kingdom is more robust now than it was in 1980, largely as a result of the controversy over four apparent suicides involving young British Army recruits at Deepcut Barracks. The presumption now (but not in 1980) is that any sudden death, even if it appeared to be a clear-cut case of accident or suicide, would be passed to the local police to investigate and not handled by the military police.

In relation to Bentwaters and Woodbridge, some officers have personal recollections of the occasional suicide at the establishment (the guesstimate of one every other year has been given), but it's a sad fact of military life that suicides in the Armed Forces do occur, not least because of the comparatively easy access to firearms. It might be expected that the problem would be most acute with young people posted away from the United States for the first time. Penniston sums it up this way: "There are, I suppose, any number of things that can happen with ten thousand people assigned to and living at the twin bases at the time. A number of crimes and things can happen at the bases, as they can happen to the general public. With that said, I have no recollections of any suicides at that base while assigned there from 1980 to 1984."

Burroughs is even more blunt: "No suicides that I'm aware of."

Notwithstanding, we have no evidence that would substantiate the claim of any suicides being directly attributable to the Rendlesham Forest incident, and neither do we have any evidence that the suicide rate at Bentwaters/Woodbridge (or among personnel posted at Bentwaters/Woodbridge at the time of the incident and subsequently posted elsewhere) was higher than might be expected when looking at suicides in the USAF as a whole.

COBRA MIST

In chapter 6 we mentioned that the military facility at Bawdsey later became home to a secret USAF research project code-named Cobra Mist, designed to develop an over-the-horizon radar system. Over the years, some people in the UFO community have tried to link Cobra Mist with Rendlesham, but it's not entirely clear that the perceived link has any substance, other than the fact that Bawdsey is geographically close to Bentwaters/Woodbridge. Exotic projects such as Cobra Mist certainly attract conspiracy theories all the time. The High-frequency Active Auroral Research Program (HAARP) is one such example. The facility studies the ionosphere with a view to enhancing communications and has spawned conspiracy theories that mainly involve weather control but also include mind control. Montauk Air Force Station on Long Island is another facility at the center of numerous conspiracy theories. These include the claim that a US Navy ship, the USS *Eldridge*, was made invisible as part of the so-called Philadelphia Experiment. They also include the allegation that the facility was the location of a secret US time travel project. Collectively, these latter allegations are known as the Montauk Project.

Cobra Mist certainly fit the pattern: a vast, spooky-looking facility, built amid great secrecy at the height of the Cold War. Construction began in 1967 and the facility was completed in 1971. But the system was plagued by a strange noise (unidentified but possibly the result of Soviet jamming) that rendered the system useless. The USAF brought in scientists from the prestigious Stanford Research Institute to help locate the problem, but without success. The project was abandoned on June 30, 1973, without ever having been formally operational. Unless this closure was a ruse (and there's no evidence that this was the case), it's difficult to see that there could be any connection with the Rendlesham Forest incident—unless one subscribes to the view that, like the Montauk Project, Cobra Mist was a facility where time travel experiments were carried out.

POSSIBLE INVOLVEMENT OF HMS *NORFOLK*

Another rumor is that between December 24 and December 30, 1980, the Royal Navy destroyer HMS *Norfolk* had been anchored off Orford Ness and that the crew had been ordered to stay belowdecks at all times, with all the ship's systems powered down. In related rumors, fishermen were allegedly told to stay away from the area between Bawdsey and Orford Ness between December 27 and December 30, while a Soviet Tu-142 maritime reconnaissance aircraft had apparently been sighted flying just off the coast.

The point about fishermen being warned away from the area might seem to corroborate the theory (mentioned briefly in chapter 9) that a nuclear leak had occurred at the nearby Sizewell A or Sizewell B nuclear power station. However, despite what conspiracy theorists believe, the UK government and the nuclear power industry tend to play a relatively straight bat when it comes to responding to nuclear leaks.

These claims were made by Brenda Butler, but we are not aware of any corroborative evidence. Indeed, in a letter dated September 3, 1987, the MoD's Naval Historical Branch responded to a query about the location of HMS *Norfolk* during the time period concerned as follows: "Between 6 December 1980 and 14 January 1981 HMS NORFOLK was at Portsmouth for an Assisted Maintenance period combined with leave."

WEATHER WEAPONS

In chapter 9 we mentioned that Rendlesham Forest had been hard hit by the Great Storm of 1987, making it difficult for anyone to compare the forest now with what it looked like in 1980. Larry Warren has claimed that when stationed at Bentwaters/Woodbridge he saw what has been dubbed a cloudbuster—a device aimed at creating rainfall. This device was based on the controversial theories of the psychoanalyst Wilhelm Reich.

It's a matter of historical record that the US and the UK governments have conducted experiments and operational missions in weather modification. The most famous US example is probably Project Popeye, in the Vietnam War. This involved seeding clouds with silver iodide, with the aim of making it rain on the Ho Chi Minh Trail, thus bogging down the main Vietcong resupply route. In the United Kingdom, RAF cloud-seeding experiments have been linked (though there is no definitive proof of this) to the Lynmouth Flood of August 15/16, 1952, in which thirty-five people died. There have also been rumors that weather modification experiments were carried out at Orfordness in the fifties. But while this technology and related programs undeniably exist, Wilhelm Reich was a maverick and his theories have been widely discredited. So even if weather modification technology was based at Bentwaters/Woodbridge, it seems difficult to believe that cloudbusters would be involved when even the dated Operation Popeye technology would have been far more effective.

This is all rather tenuous: a mixture of historical rumor and fact about weather modification experiments, most of which predates the Rendlesham Forest incident; a claim by Larry Warren; and the fact that (years after the Rendlesham Forest incident) Rendlesham Forest was hard hit by a freak storm.

UNUSUAL TREE FELLING

There is a rumor that a large number of trees were felled after the incident, perhaps because they had been irradiated. Undeniably, many trees were felled by the Great Storm of 1987 and a number of damaged ones were cut down immediately thereafter. Aside from this, the Forestry Commission felled trees from time to time as part of normal forestry procedures (e.g., if a tree became diseased or was damaged in a storm) and the USAF had an interest in ensuring that trees were felled around the twin bases, so they couldn't be approached covertly. Lord Hill-Norton asked a PQ about this and the exchange is recorded in the *Hansard* dated April 30, 2001:

> **Lord Hill-Norton asked Her Majesty's Government:**
>
> *Whether they requested or instructed the Forestry Commission to fell any trees in Rendlesham Forest or Tangham Woods in the aftermath of the Rendlesham Forest incident; and, if so, on what grounds.*
>
> **Baroness Hayman** *The Forestry Commission was not instructed to fell any trees after the alleged incident in Rendlesham Forest in December 1980. Most of the trees in the area had been selected and marked for felling well before the alleged incident and were felled several months after it.*

This answer seems to confirm that there was a noticeable tree felling in the aftermath of the incident, but that it was routine and pre-planned and did not take place until some months after the incident.

WERE LOCAL PRISONS ON EVACUATION ALERT?

One of the most bizarre stories in relation to the Rendlesham Forest incident comes from retired prison officer George Wild. Wild was a senior prison officer at HM Prison Leeds (also known as Armley Gaol) and some time before his retirement attended a seminar with other prison officers at which the evacuation story was relayed in the margins of the formal business. An officer based at Highpoint Prison in Suffolk told Wild that on December 27, 1980, the staff at the prison were put on alert and told to prepare for a possible evacuation. No further details were given, aside from a vague statement that the reason for this was "national security." In the event, no evacuation took place.

Wild later confirmed that not only Highpoint Prison but also Hollesley Bay Youth Correction Centre had received a warning that an evacuation might be necessary. Nor was Wild the only source of this story. Brenda Butler had heard from a local police officer about the potential evacuation of the Hollesley Bay facility and had also received a letter from a prisoner at a third local institution, Blundeston Prison. This letter referred to a UFO sighting on December 27 (which would tie in with the second night's sighting involving Halt and his men) and also mentioned the fact that the prison was on standby alert, though the details were not given and it wasn't clear whether or not an evacuation had been ordered. Despite a promise to contact Butler again on his release, nothing further was heard from the individual concerned, and one might reasonably suppose that a prisoner was less likely to know whether or not an evacuation had been ordered than would a prison officer or a local police officer.

Lord Hill-Norton decided to probe this issue by asking a series of written questions in Parliament. The initial question and answer, on October 23, 1997, read as follows:

Lord Hill-Norton asked Her Majesty's Government:

Whether staff at Highpoint Prison in Suffolk received instructions to prepare for a possible evacuation of the prison at some time between 25 and 30 December 1980, and if so, why these instructions were issued.

Lord Williams of Mostyn I regret to advise the noble Lord that I am unable to answer his Question, as records for Highpoint Prison relating to the period concerned are no longer available. The governor's journal is the record in which a written note is made of significant events concerning the establishment on a daily basis. It has not proved possible to locate that journal.

Lord Hill-Norton returned to this subject on January 23, 2001, probing the situation with regard to the two other local prisons that had been mentioned:

Lord Hill-Norton asked Her Majesty's Government:

Whether staff at Blundeston Prison or Hollesley Bay Youth Correction Centre received any instructions to prepare for a possible evacuation at some time between 25 and 30 December 1980; and if so, why these instructions were issued.

The Parliamentary Under-Secretary of State, Home Office (Lord Bassam of Brighton) *We can find no record of any such instructions.*

There was another question on April 26, 2001, in which Lord Hill-Norton probed the issue of the governor's journals for the locations concerned:

Lord Hill-Norton asked Her Majesty's Government:

Further to the Written Answer by Lord Bassam of Brighton on 23 January (WA 8), whether their search for evidence of any instructions concerning the possible evacuation of Blundeston Prison and Hollesley Bay Young Offender Institution included an examination of the governor's journals for these two establishments; and whether these journals have been retained.

The Parliamentary Under-Secretary of State, Home Office (Lord Bassam of Brighton) *Governors' journals are the most likely source of this information so long after the event. The governor's journal at Blundeston remains in existence and was examined. The relevant governor's journal for Hollesley Bay could not now be found, and in the absence of any other written record, long-serving staff, including the governor's secretary, were consulted. They did not recall any instruction to prepare for an evacuation although they well remembered the local events of the time which prompted speculation about such an instruction.*

There were two final questions on October 4, 2001:

Lord Hill-Norton asked Her Majesty's Government:

Further to the Written Answer by Lord Bassam of Brighton on 26 April (WA 240), whether the examination of the governor's journal at Blundeston prison revealed any details of an alert during 25 to 30 December 1980; and whether in this period there was any mention of RAF Bentwaters, RAF Woodbridge or Rendlesham Forest.

Lord Rooker The governor's journal revealed no such details and there was no mention of RAF Bentwaters, RAF Woodbridge or Rendlesham Forest.

Lord Hill-Norton asked Her Majesty's Government:

What is their response to the absence of the governor's journals covering the period 25 to 30 December 1980 in respect of Hollesley and Highpoint prisons; and whether, in the absence of these records, they will consult the then governors about any alert or warning to evacuate during that period.

Lord Rooker The governor's journal is a record of day-to-day events, and the absence of journals so long after the event is not a cause for concern. The governors of Hollesley Bay and Highpoint prisons in December 1980 are no longer in the service and I am not persuaded that the effort required to trace them is justified.

The fact that governors' logs that might just have given a clue as to what happened had gone missing was suspicious to Lord Hill-Norton and reminiscent of another UFO incident in which Lord Hill-Norton had taken an interest. It is alleged that on an unknown date in 1999 several hundred personnel on British and Norwegian warships taking part in a NATO exercise sighted a UFO. It's further claimed that one of the ships tracked the UFO on radar. Lord Hill-Norton wrote to one of the ministers at the MoD on September 24, 2002, asking that the ship's log be searched for reference to the UFO encounter.

The reply that Lord Hill-Norton received was dated October 21 and read as follows:

You asked that HMS MANCHESTER's log for the periods 26 October to 6 November 1998 and 8 February to 3 March 1999 be scrutinized for references to unidentified aerial craft sighted by the ship's company. No such references have been found in any of the log entries which are available.

Unfortunately, I have to add the rider that HMS MANCHESTER's log covering the period 1 Feb until sunrise on 13 February 1999 was lost in Bodo, Norway, during the deployment. The log was positioned, as is the custom, at the head of the gangway when the vessel was alongside in port, and an unusually strong gust of wind carried it overboard. The circumstances are properly recorded and certified by HMS MANCHESTER's Commanding Officer in the log opened on 13 February following this loss. In light of the missing document, my officials have contacted the Commanding Officer of the MANCHESTER at the time. He has stated that nothing which could be remotely construed as an unusual event or sighting involving unidentified aerial craft occurred during this or any other of MANCHESTER's deployments while he was in command.

Lord Hill-Norton was astounded by this response, and his October 29, 2002, reply, dripping with sarcasm, read as follows:

Thank you for your letter of Trafalgar Day (did you know? Did any of your officials?).

I am grateful for the trouble you, and they, have taken, we shall now return to our informant, a member of the ship's company at the relevant time, and press him further. I am also in touch with Norwegian Naval people about the incident.

I have to tell you that in 49 years in the Royal Navy, which included more than 30 years at sea in more than 25 different ships, I have never heard of a rough deck log being blown over the side, more particularly in harbor (we do not say port in the Royal Navy—two new pieces of information in one letter).

Some less charitable persons than myself, might even consider it odd that this unique occurrence should have surrounded a perfectly legitimate enquiry about UFOs. Even my charitable mind finds that credulity is thereby strained pretty close to the limit. I hope your own has stood the remarkable strain so well.

I will return to the charge after we have been in touch with our eye-witness.

In the meantime your letter will be an unusual and perhaps useful addition to the dossier we are compiling.

Setting aside Lord Hill-Norton's experience of the disappearing ship's log and governors' journals, what could the significance be of the possible evacuation of the local prisons, assuming that this story is correct? Two scenarios come to mind. The first possibility is that the authorities were aware of something that might pose a danger to people at the prison(s) and put them on standby in case this danger materialized. However, if this was the case, one would have expected other locations in the area (e.g., hospitals and private residences) to be included, too. The second and far more frightening possibility is that secure locations were required for people who might want to leave the area against the wishes of the authorities. Running through a scenario analysis, the sort of situation that springs to mind is where large numbers of people (the capacity of Highpoint Prison is around thirteen hundred) had been irradiated or exposed to a chemical or biological agent.

BIOLOGICAL HAZARD

When undertaking hypothetical discussions about the extraterrestrial hypothesis few issues are more chilling than the potential biohazard. The issue had first arisen in 1960 when the Space Science Board advised NASA that a quarantine procedure should be established to ensure that any spacecraft and samples returned to Earth following space missions were free of any organisms that might threaten the Earth's biosphere. NASA uses the phrase "back contamination" to describe this, and the issue really came into focus in the run-up to the Apollo 11 moon mission. The Space Science Board had summed up the issue in this way: "The introduction into the Earth's biosphere of destructive alien organisms could be a disaster . . . we can conceive of no more tragically ironic consequence of our search for extraterrestrial life."

A more detailed Space Science Board statement on the issue read as follows:

> *The existence of life on the moon or planets cannot . . . rationally be precluded. At the very least, present evidence is not inconsistent with its presence . . . Negative data will not prove that extraterrestrial life does not exist; they will merely mean that it has not been found. To contain any alien life forms, astronauts, spacecraft, and lunar materials coming back from the moon should be placed immediately in an isolation unit; the astronauts should be held in rigid quarantine for at least three weeks; and preliminary examination of the samples should be conducted behind absolute biological barriers, under rigid bacterial and chemical isolation.*

In the event, the Lunar Receiving Laboratory was constructed in order to ensure that a secure quarantine facility existed for lunar samples and for the astronauts themselves.

The issue remains a complicated one to this day, first because of the question of whether alien life exists in the first place (self-evidently, if you think it doesn't no risk arises) and second because even if it does, worrying about a threat from alien microbes makes one sound as if one has been watching too much science fiction. The scientific debate about whether extraterrestrial life could ever pose a threat to Earth's biosphere is beyond the scope of this book, but in the United States the issue is the responsibility of NASA's Office of Planetary Protection. The United Nations has responsibilities here, too, and the UN's Committee on Space Research (COSPAR) has a Panel on Planetary Protection, as well as a formal COSPAR Planetary Protection Policy.

In the absence of any formal contingency plan to deal with a potential biohazard resulting from a UFO incident, the COSPAR and NASA Planetary Protection Policy would be little use, because it consists largely of preventive measures. Of more use would be existing plans for a chemical or biological incident (e.g., a terrorist attack or an inadvertent leak) or the Home Office document "Satellite Accidents" (of which the MoD's UFO project was aware), which includes consideration of the measures to be taken to deal with the possible radiation hazard in the event of the crash of a nuclear-powered satellite.

In the United Kingdom, the key establishment in the event of dealing with any biological hazard resulting from extraterrestrial organisms of any kind would doubtless be the Defence Science and Technology Laboratory at Porton Down. Intriguingly, this establishment features in a separate rumor about the Rendlesham Forest incident.

The evidence for this Porton Down connection is inconclusive. On the plus side, it comes from author and investigative journalist Georgina Bruni, who had strong links with the MoD. On the minus side, Bruni's sources for this story were apparently RAF Police officers whose names are not known—Bruni declined to share their identities, on the basis that one was still serving and was wary of repercussions.

The story from the RAF Police involves the nearby base at RAF Watton, which was involved, as readers will recall, in the radar tracking of at least one of the UFO incidents. On 27/28 December (the same night as Halt's encounter) two RAF Police dog handlers were tasked to investigate strange lights just beyond the perimeter fence. At the same time, allegedly, an uncorrelated target was being tracked on radar. The precise details of this radar report are not known, but RAF Watton is around forty miles from Bentwaters/Woodbridge, making a Rendlesham connection possible. The dog handlers were astounded to find several figures shining lights into the sky. The figures were wearing clothing that looked like an NBC (Nuclear, Biological, Chemical) suit. One of the witnesses stated that they were subsequently interviewed by a senior officer, who advised them to forget what they had seen, as it had just been poachers—an explanation that wasn't believed. The witness also stated that their police notebooks were confiscated and that various logbooks in which the incident may have been recorded subsequently went missing. Readers will doubtless have noticed the ever-increasing list of logbooks, ships' logs, governors' logs, and notebooks that go missing in the aftermath of UFO incidents!

All this may tie in with Jim Penniston's recollection that a CIA "containment study team" was sent out to examine the landing site in Rendlesham

Forest. Georgina Bruni had other sources who confirmed that a study team of some sort was sent out to check the site, but it's not entirely clear whether this was a US or a UK team. And in all of this, it's also not clear whether any such team was checking for a radiological hazard (as might be the logical first assumption, given that Halt's report to the MoD specifically mentioned radiation readings taken at the landing site) or a biological hazard.

If personnel from Porton Down *did* visit the forest, their first priority would be to see if any biological agent was present. If anything unusual was detected, options would have included destroying it. It is more likely, given the potential implications in terms of biotechnology, including bioweapon, research, that containment and removal would be the preferred option. Again, however, the same study team would doubtless be equipped with means of assessing not just a biohazard but a radiological hazard.

Lord Hill-Norton probed this issue by asking another PQ. The exchange was printed in *Hansard* dated January 25, 2001:

Lord Hill-Norton *asked Her Majesty's Government:*

Whether personnel from Porton Down visited Rendlesham Forest or the area surrounding RAF Walton in December 1980 or January 1981; and whether they are aware of any tests carried out in either of those two areas aimed at assessing any nuclear, biological or chemical hazard.

Baroness Symons of Vernham Dean *The staff at the Defence Evaluation and Research Agency (DERA) Chemical and Biological Defence (CBD) laboratories at Porton Down have made a thorough search of their archives and have found no record of any such visits.*

I should declare an interest here. Aside from my own service with the MoD, my father served—at considerably higher level—in the MoD for many years, his numerous posts including Deputy Director (Weapons), Assistant Chief Scientific Adviser (Projects), Deputy Controller and Adviser (Research and Technology) and Deputy Chief Scientific Adviser. He was also awarded the US Secretary of Defense Medal for Outstanding Public Service. At more than one stage in his career he had responsibility for the Porton Down facility and he never once even hinted that the establishment had a role in the Rendlesham Forest incident.

MEN IN BLACK (MIB)

Following on from the preceding, there has been a suggestion that in the immediate aftermath of the Rendlesham Forest incident smartly suited men were searching the forest and/or quizzing local people about what they had seen. These rumors might be dismissed as fantasy or exaggeration, were it not for the fact that one of the witnesses is the man most associated with the skeptical theory that the witnesses misidentified the Orfordness Lighthouse, local forester Vince Thurkettle. Discussing this with Georgina Bruni in around 1999, he described how he encountered two smartly dressed men very shortly after the UFO sightings, sometime between December 29 and New Year's Day. Thurkettle recalled that the smart clothing was decidedly inappropriate for the forest. The men did not identify themselves but asked whether Thurkettle had been out in the forest at any time over the previous four nights. The men were not in uniform and Thurkettle did not say whether they had American or British accents. At first he wondered whether they were reporters, but no story appeared and we have already seen that the media were not aware of the UFO sightings until much later. It's just possible that these were reporters who never found a story solid enough to run, but it's equally possible that these men were from the USAF or from the RAF or the MoD—perhaps associated in some way with Porton Down.

Long before the famous movie franchise, MIB have been a familiar trope in ufology. In a variety of cases, these sinister, mysterious, smartly dressed figures have visited witnesses in the aftermath of UFO sightings, quizzing them about their encounters and ordering them not to discuss what they saw. The consensus among believers is that these individuals are from the government—the image is that of the G-man. But on the basis of some cases where truly bizarre behavior (such as apparent unfamiliarity with everyday objects, and odd speech) has been reported, others have speculated that these strange individuals are extraterrestrials, presumably engaged in some strategy to cover their own tracks. Other more prosaic theories are that they are journalists after a story, people from local UFO groups, practical jokers, or "Walter Mitty" type fantasists. Some MIB reports, of course, will be hoaxes themselves.

We have no reason to doubt the testimony of Vince Thurkettle, and indeed a number of other local people tell similar stories, though in these other instances the rumor is that this happened on New Year's Day itself, whereas Vince believes it was closer to December 29.

In this chapter we addressed some miscellaneous rumors about the Rendlesham Forest incident that did not fit into the story narrative and were best addressed as stand-alone issues. In looking in particular at rumors concerning a potential biological or radiological hazard posed by the events, we have raised a frightening possibility about the incident that has immense implications, in particular for the witnesses most closely involved, namely, John Burroughs and Jim Penniston.

13. NO DEFENSE SIGNIFICANCE?

The MoD's UFO project has featured quite a lot in this story already, and while we've pointed out some similarities between the UK work and earlier USAF programs such as Project Blue Book, it's appropriate to go into more detail about the history and day-to-day work of this project. In particular, we want to show how the MoD consistently downplayed its interest and involvement with UFO research and investigation and go on to illustrate how this was done with specific relation to the Rendlesham Forest incident. First, we need to show how it all began.

The UK government had been aware of two separate mysteries involving strange objects in the sky that pre-dated 1947, when the US government first looked at the subject. The first were "Foo Fighters"—balls of light, or in some cases metallic craft, that appeared to follow RAF aircraft during bombing raids over Nazi territory during the Second World War. At first, it was thought that these were some sort of Axis secret weapon, but there was no apparent hostility. After the war, it transpired that some German and Japanese pilots had seen similar things and concluded that they were Allied weapons of some sort. The second were the so-called ghost rockets. These mystery objects were seen over parts of Scandinavia in 1946. Both the US and the UK governments took an interest in this, believing the objects concerned were prototype Soviet rockets of some sort, back-engineered from the Nazi V-1 and V-2 rockets in the same way as the US authorities were building on this technology as part of the research effort that would lead directly to the Apollo moon rockets.

The key figure in the establishment of the MoD's UFO project was Sir Henry Tizard. Tizard is best known for his pioneering work on the development of radar technology prior to the Second World War, and his various wartime posts included Scientific Advisor to the Air Staff. He returned to the MoD in 1948 as Chief Scientific Advisor, a post that he held until 1952.

Tizard had very close links with the US government. During the Second World War, he led what became known as the Tizard Mission: an exchange of information and technology between the United States and the United Kingdom. Tizard's main contact was the Chairman of the National Defense Research Committee, Vannevar Bush, and the main areas involved were radar, the jet engine, and nuclear research.

In 1950, Tizard had become intrigued by the increasing media coverage of UFO sightings in the United States, the United Kingdom, and other parts of the world. Using his authority as Chief Scientific Advisor at the MoD, he decided that the subject should not be dismissed without some proper, official investigation. Accordingly, he agreed that a small Directorate of Scientific Intelligence/Joint Technical Intelligence Committee (DSI/JTIC) working party should be set up to investigate the phenomenon. This was dubbed the Flying Saucer Working Party. The terms of reference of the Flying Saucer Working Party read as follows:

1. *To review the available evidence in reports of "Flying Saucers."*

2. *To examine from now on the evidence on which reports of British origin of phenomena attributed to "Flying Saucers" are based.*

3. *To report to DSI/JTIC as necessary.*

4. *To keep in touch with American occurrences and evaluation of such.*

The five-man working party was chaired by Mr. G. L. Turney from one of the MoD's scientific intelligence branches. All the members were specialists in the field of scientific and technical intelligence.

The working party's conclusions were set out in a document dated

June 1951 and bearing the designation DSI/JTIC Report No. 7. It was titled "Unidentified Flying Objects" and classified "Secret Discreet." The report comprised six pages (including the cover sheet) and concluded that all UFO sightings could be explained as misidentifications of ordinary objects or phenomena, optical illusions, psychological delusions, or hoaxes. The main body of the report ended with the following statement: "We accordingly recommend very strongly that no further investigation of reported mysterious aerial phenomena be undertaken, unless and until some material evidence becomes available." The report was duly considered by the DSI/JTIC, and Mr. Turney recommended that in view of its skeptical conclusions it should be regarded as a final report. He further suggested that the working party be dissolved with immediate effect. This was agreed by the meeting, thus bringing to an end the MoD's first UFO research project.

Just a few months later, in 1952, there was a massive wave of UFO sightings over the United Kingdom, where many of the witnesses were RAF personnel and where some of the visual sightings were corroborated by radar. Many of these sightings had occurred in the margins of Operation Mainbrace, a NATO military exercise that was the largest exercise held since the Second World War. As a result of this, the skeptical conclusions of the Flying Saucer Working Party were overturned and in 1953 the MoD set up a process whereby UFO sightings reported to the department would be investigated.

From the very outset, UK policy in relation to UFOs was influenced by the Americans. As mentioned earlier, the MoD's Flying Saucer Working Party had been given the remit of liaising with those US authorities undertaking research and investigation into the UFO phenomenon. Once the terms of reference included a requirement to get alongside the Americans on the UFO issue, active liaison began. A member of the Flying Saucer Working Party had traveled to America to meet with US authorities.

One of the people consulted by the British was H. Marshall Chadwell, who sat in on at least one of the Flying Saucer Working Party's meetings. Chadwell was Assistant Director of the CIA's Office of Scientific Intelligence and in 1952 and 1953 was one of the key figures in one of the US government's best-known studies into UFOs, the scientific panel on UFOs— better known as the Robertson Panel, after its chairman, H. P. Robertson, an eminent physicist from the California Institute of Technology.

More proof of the US influence comes from the testimony of Edward Ruppelt, one of the former heads of the USAF's Project Blue Book. Writing in his 1956 book, *The Report on Unidentified Flying Objects,* he stated: "Two RAF intelligence officers who were in the US on a classified mission brought six single-spaced typed pages of questions they and their friends wanted answered." Ruppelt also gives a telling insight into the Operation Mainbrace sightings, mentioned earlier, that led to the United Kingdom setting up its own formally constituted UFO project: "It was these sightings, I was told by an RAF exchange intelligence officer in the Pentagon, that caused the RAF to officially recognize the UFO." Proof of the US influence can be seen clearly in the Flying Saucer Working Party's final report, which began with a discussion of the US government's UFO project. There was also mention of some well-known American UFO sightings.

When the United Kingdom started formally investigating UFO sightings in 1953, the US influence was overwhelming. Not only were the United Kingdom's investigative terms of reference and methodology a mirror image of those of Project Blue Book, but even the forms used to record the details of a sighting were virtually identical. So, too, was the way in which the British government downplayed the phenomenon (and the extent of official interest) with the media and the public.

The MoD's UFO project ran from 1953 to 2009. In that time, it was embedded within various different MoD divisions, some military, others civilian. The list includes DDI(Tech), S4, S6, DS8, Sec(AS), and DAS. The fact that so many sections have been involved with this subject over the years has spawned conspiracy theories about there being a myriad of different divisions involved in UFO investigations, whereas the reality of the situation is that the different acronyms and abbreviations often simply reflected MoD reorganizations. For this reason, the media and the public tend to use descriptive terms such as "MoD's UFO project" or "the UFO desk" to describe the work. All this could have been avoided had the MoD decided to give this work a formal name, but to be fair, even the United States went through three project names (Sign, Grudge, and Blue Book), showing that governments seldom make things simple for people to understand.

I joined the MoD in 1985 and resigned in 2006, after a successful, interesting, and enjoyable twenty-one-year career that culminated in my serving as an acting Deputy Director in the Directorate of Defence Security.

From 1991 to 1994 I ran the UFO project, which at the time was embedded in Sec(AS). A very brief public acknowledgment of this section (and my work in it) was given as a result of a PQ asked by Norman Baker, a Member of Parliament who champions open government. Baker's question was answered by the Under Secretary of State for Defence, Don Touhig, and is recorded in the *Hansard* dated April 18, 2006. The exchange reads as follows:

> To ask the Secretary of State for Defence in what capacity Mr Nick Pope was employed by his Department between 1991 and 1994.
>
> From 1991 to 1994 Mr Pope worked as a civil servant within Secretariat (Air Staff). He undertook a wide range of secretariat tasks relating to central policy, political and parliamentary aspects of non-operational RAF activity. Part of his duties related to the investigation of unidentified aerial phenomena reported to the Department to see if they had any defence significance.

When I took over this position in 1991, we used to receive two to three hundred reports each year. The methodology of an investigation was fairly standard. First, you interviewed the witness to obtain as much information as possible about the sighting: date, time, and location of the sighting, description of the object, its estimated speed, height, et cetera. Then you attempted to correlate the sighting with known aerial activity such as civil flights, military exercises, or weather balloon launches. We could check with the Royal Observatory Greenwich, to see if astronomical phenomena such as meteors or fireballs might explain what was seen. We could check to see whether any UFOs seen visually had been tracked on radar. If we had a photograph or video, we could get various MoD specialists to enhance and analyze the imagery. We could also liaise with staff at the Ballistic Missile Early Warning System at RAF Fylingdales, where they have space-tracking radar. Finally, on various scientific and technical issues we could liaise with the DIS, though this is an area that I am not able to discuss in any detail, due to the sensitivities.

After investigation, around 80 percent of UFO sightings could be explained as misidentifications of something ordinary, such as aircraft lights, satellites, airships, weather balloons, meteors, or bright stars and planets. In around 15 percent of cases there was insufficient information to draw any firm conclusions. Approximately 5 percent of sightings seemed to defy any conventional explanation.

The cases that were of most interest to us were UFO sightings where there were multiple witnesses or where the witnesses were trained observers such as police officers or military personnel; sightings from civil or military pilots; sightings backed up by photographic or video evidence, where technical analysis found no signs of fakery; sightings tracked on radar; and sightings involving structured craft seemingly capable of speeds and maneuvers that even the most advanced aircraft and drones were not able to match.

The fact that the MoD was engaged in this work was not a secret. Indeed, a good deal of the job involved dealing with the public, Parliament, and the media. It could hardly be any other way: statistically, most UFO sightings came from the public, and clearly we had to deal with the public to get this data! However, it was an unequal partnership: the MoD wanted to glean as much information as possible from witnesses, while offering as little as possible in return. There was, in short, a deliberate policy to downplay the true extent of our involvement and interest in the subject.

The best way to illustrate this is to examine the policy statement that appeared on the MoD's Web site up until the 2009 axing of the UFO project. It appeared under the heading "MoD Policy on Unidentified Flying Objects (UFO)" and read as follows:

> The Ministry of Defence does not have any expertise or role in respect of "UFO/flying saucer" matters or to the question of the existence or otherwise of extraterrestrial lifeforms, about which it remains totally open-minded. To date the MoD knows of no evidence which substantiates the existence of these alleged phenomena.
>
> The MoD examines any "UFO" reports it receives solely to establish whether what was seen might have some defence significance; namely, whether there is any evidence that the United Kingdom's airspace might have been compromised by hostile or unauthorized air activity. Unless there is evidence of a potential threat to the United Kingdom from an external source, and to date no "UFO" report has revealed such evidence, we do not attempt to identify the precise nature of each

sighting reported to us. We believe that rational explanations, such as aircraft lights or natural phenomena, could be found for them if resources were diverted for this purpose, but it is not the function of the MoD to provide this kind of aerial identification service. It would be an inappropriate use of defence resources if we were to do so.

If you wish to report a sighting or have any questions about the MoD's position regarding UFOs, you should write to the following address:

> *Ministry of Defence*
> *Directorate of Air Staff—Freedom of Information*
> *5th Floor, Zone H*
> *Main Building*
> *London*
> *SW1A 2HB*

Alternatively you can contact us on any of the following;

> *Telephone: 020-7218-2140 (24 hour Answerphone)*
> *Fax: 020-7218-7701 or 020-7218-2680*
> *E-Mail: das-ufo-office@mod.uk*

This statement was carefully crafted, and indeed I drafted many similar such letters myself, during my time on the MoD's UFO project. It wasn't an out-and-out falsehood, but it was misleading because it really only told half the story. It was certainly true to say that we believed *most* (but not all) UFO sightings had conventional explanations. It was also true to say—despite what conspiracy theorists believed—that we had no definitive proof that any UFO sightings were extraterrestrial. But to say the MoD had no expertise or role in UFO matters was farcical, given that the department had been researching and investigating UFO sightings since 1953. But if you look carefully, you'll notice that the statement says "UFO," not UFO—and goes on to use the outdated term "flying saucer." In other words, the MoD was deliberately using the term "UFO" in the erroneous, pop culture way where the term is synonymous with extraterrestrial spacecraft. In that context, of course, the claim was true: MoD has no role or experience with regard to aliens!

Use of the phrase "flying saucer" is interesting. The term was used up until the early fifties, when it was replaced by the term "Unidentified Flying Object." This latter term was devised by staff on the USAF's Project Blue Book and was soon adopted by the MoD but also by the media and the public. More to the point, the term "flying saucer" is now archaic and has negative connotations in the same way that the term "UFO nuts" is sometimes used. The MoD has similarly used phrases such as "little green men" in relation to the subject when responding to requests from the media for statements on UFO-related stories. Again, the term—at least in UK culture—is archaic and disparaging, designed to belittle the subject and make it comedic. Another example of this policy is use of the phrase "UFO spotters" in relation to people who study the phenomenon. The term "spotters" is disparaging and the clear implication is "nerds."

Later in the MoD's statement the phrase "We believe that rational explanations, such as aircraft lights or natural phenomena, could be found for them" appears. This is a clever piece of business. When I was running the UFO project I used a phrase along the lines of "We believe that *conventional* explanations could be found for most UFO sightings." Use of the word "rational" in the new version implies that anyone who thinks otherwise is irrational.

The key phrase in the MoD statement is the term "defence significance," which appears as part of this wider phrase: "The MoD examines any 'UFO' reports it receives solely to establish whether what was seen might have some defence significance." Over the years, the term "defence interest" has also been used. Usually, during my time on the UFO project, there was no attempt to define it—a deliberate tactic to keep things vague. Interestingly, the statement quoted here does attempt a definition: "namely, whether there is any evidence that the United Kingdom's airspace might have been compromised by hostile or unauthorized air activity." The word "airspace," followed quickly by the phrase "air activity," clearly implies that the sole interest here is foreign military aircraft. That's certainly one of the key areas of interest, and the possibility that some UFO sightings might turn out to be foreign military aircraft probing our airspace was obviously one reason why the MoD investigated these sightings. Indeed, I personally coined a phrase along the lines of "more likely Russian than Martian," which had the double effect of focusing the debate on conventional aircraft and of making the whole subject seem humorous.

In relation to the judgment as to whether UFOs are of any "defence significance," the MoD's statement goes on to say this: "Unless there is evidence of a potential threat to the United Kingdom from an external source, and to date no 'UFO' report has revealed such evidence"—an assessment that has sometimes been given in the form of stating that UFOs are adjudged to be of "no defence significance." But if the "no defence significance" assessment depends upon there being no evidence of "a potential threat to the United Kingdom from an external source," then the MoD is on dangerous ground. Because while the absence of hostile action might justify saying there's no direct threat, there's self-evidently a "potential threat" if there's *any* evidence that objects of unknown origin have penetrated the United Kingdom's Air Defence Region. And as the MoD's own files show, there have been plenty of UFO sightings (most notably the Rendlesham Forest incident) where there's clear evidence that this is *precisely* what happened. So the MoD's public statement on UFOs can be shown to be false, by simple application of logic. Even the assessment of there being no hostile action might be wishful thinking, as opposed to being a proper threat assessment in the sense that, say, an intelligence officer specializing in counter-terrorism would mean it.

The Rendlesham Forest incident posed a special challenge for the MoD in terms of downplaying both the sighting itself and the extent of the department's involvement. We saw in chapter 11 how the MoD responded in the immediate aftermath of the *News of the World* breaking the story, with a low-key press line, the key parts of which read as follows:

I can confirm that the Ministry of Defence did receive a report from base personnel of a UFO sighting near RAF Woodbridge on 27 December 1980. . . . The report was dealt with in accordance with normal procedures, i.e., it was passed to staff concerned with air defence matters who examine such reports to satisfy themselves that there are no defence implications. In this instance MoD was satisfied that there was nothing of defence interest in the alleged sightings.

This low-key rendering of the events was not exactly a falsehood, but it was at best a selective and misleading account, ignoring the most compelling parts of the story. The "nothing of defence interest in the alleged sightings" is the standard line on UFOs but is utterly at odds with the facts here. Use of the word "alleged" is churlish at best and an insult to the numerous USAF witnesses at worst. But aside from this immediate reaction to the *News of the World* exclusive, how did the MoD spin the Rendlesham Forest incident in later years?

The standard reply that members of the public would receive from 1983 onwards was usually the following. It was very similar to the version used in the press line, quoted earlier, but had evolved over time, so as to be even more dismissive:

> *I can confirm the Ministry of Defence did receive a report from base personnel of a UFO sighting near RAF Woodbridge on 27th December 1980. The report was dealt with in accordance with normal proce-dures, i.e. it was passed to staff concerned with air defence matters who examine such reports to satisfy themselves that there are no de-fence implications. In this instance MoD was satisfied that there was nothing of defence interest in the alleged sightings. There was no ques-tion of any contact with "alien beings" nor was there any confirmation that an object had landed in the forest.*

The addition of the final sentence is interesting. By mentioning "alien beings" it focused on the uncorroborated rumors, despite the fact that none of the main witnesses had reported having seen anything other than an unidentified craft. The effect, whether intentional or not, was to spin the whole affair as a wild-sounding tale involving aliens, as opposed to a series of incidents in which numerous witnesses, including the Deputy Base Commander, saw an unidentified craft.

But there followed another paragraph that put a very different spin on things:

> *You may be interested to know that the BBC recently carried out its own investigations into the incident and concluded that the UFO was*

nothing more sinister than the pulsating light of the Orfordness Light-house some 6 or 7 miles away through the trees.

This statement was stretching the truth, to say the least. The BBC had certainly run a news feature on the Rendlesham Forest incident, in which Vince Thurkettle's lighthouse theory had featured, but it was hardly a definitive investigation or conclusion. Respected institution though the BBC is, their report effectively just looked at Halt's one-page memo and Vince Thurkettle's theory. The BBC had no access to any of the other documents or witness statements, let alone the witnesses themselves.

In strictness, the MoD should not have been taking sides in the debate about UFOs, but while arguably not endorsing a particular theory, even highlighting one favors it, by simple virtue of drawing attention to it. While not explicitly supporting the theory that the witnesses misidentified the lighthouse, the MoD was certainly implying as much. And as we've seen, the lighthouse theory is far from being the most convincing of the skeptical theories, with several others being more plausible and better fitting the facts. But the lighthouse theory has just the right amount of "giggle factor" to effectively make the whole story seem faintly ridiculous. Thus, one can well understand why the MoD chose to highlight it when questions were asked.

Penniston and Burroughs are scathing about the MoD's "no defence significance" sound bite and about the sorts of people who parrot the MoD's party line without any critical thought:

> *The MoD had their own brainstorm of damage control: ignore it and it goes away (at least publically). The MoD had a whole cache of "useful idiots" in the debunkers. All in all, the debunkers were targeting the cover story (containment story) and eventually it goes down in the annals of UFO folklore. The worst mistake the MoD made was saying it was of "no defence significance." That showed the event happened. The key for Pandora's Box laid with the MoD and the tons of files and communication in their archives.*

When I joined the UFO project in 1991, the department was still getting a steady stream of questions about the UFO encounter in Rendlesham

Forest from parliamentarians, the media, and the public. It was, after all, regarded as the United Kingdom's most significant UFO encounter and indeed had been dubbed Britain's Roswell. I would receive letters asking what had happened, what we knew, and what we thought. Some of the letters were polite requests for answers; others were more accusatory, implying (or directly stating) that we were covering up the truth about what had happened. Either way, all letters had to be answered.

I pretty much followed the line that my predecessors had taken, but there was another problem that was becoming increasingly difficult to manage, in relation to both spinning the Rendlesham Forest incident and downplaying the UFO subject more generally. An integral part of this was not to acknowledge the role of the DIS in relation to UFO research and investigation. This would have been relatively straightforward were it not for an administrative error mentioned previously, whereby a photocopy of a UFO report had been sent to a ufologist with the distribution list left on, instead of blacked out, as was usual—or, to be precise, whited out, so that it was not apparent that there was a distribution list at all. Another sleight of hand, done on the basis that a distribution list of several other sections would nail the lie that this was a subject we didn't take seriously! The upshot of the clerical error was that by the early nineties most people in the UFO community were aware that Defence Intelligence was involved in the MoD's UFO work.

"Intelligence" seems to be a magic word with the media and the public. Clearly it's a fascinating area, and having worked alongside various intelligence personnel in my twenty-one-year MoD career, I can testify that the work can certainly be exciting. But it's not always glamorous and it certainly doesn't quite live up to the reputation that James Bond films and movies about the CIA might imply. However, having told ufologists for years that we regarded their subject as being of "no defence significance" but then to have them find out that the DIS was involved with the subject was a recipe for disaster. It was only a matter of time before the MoD's defensive line on UFOs and on the Rendlesham Forest incident crumbled. The reckoning came in 2001.

While the UK didn't get a Freedom of Information Act until 2000 (which came fully into force in 2005), in 1994 the Conservative government led by Prime Minister John Major introduced a piece of legislation called the Code of Practice on Access to Government Information. It's best described as a watered-down, prototype Freedom of Information Act. People could request information and government departments had to release it if they held it, but the requests had to be "reasonable" and the exemptions were extremely broad, especially where defense and national security were concerned. But it was a step up from the old Public Records Act, which essentially ensured that any government information would be withheld for thirty years before it could even be considered for release!

It's surprising that the UFO community did not exploit the Code of Practice sooner than they did or more skillfully. One person who did was Georgina Bruni.

Georgina Bruni led a remarkable life and worked variously as a go-go dancer, fashion designer, and nightclub manager. She traveled extensively and at various times lived in Jersey, Italy, Hong Kong, and America before settling in London in 1992. She wrote poetry, designed a positive-thinking course, and founded one of the United Kingdom's first online magazines, called *Hot Gossip*. She was a former director of the Yacht Club, where she was involved in hosting social events for MPs, diplomats, and MoD officials. Later on she became a PR consultant and ran a social club, Le Club 2000. It was through her involvement with the Yacht Club that she first began to mix with various MoD, military, intelligence, and diplomatic staff.

In early 2001 Bruni used the Code of Practice on Access to Government Information to apply for all documents that the MoD had on the Rendlesham Forest incident. She did this in conjunction with former Chief of the Defence Staff Lord Hill-Norton, with whom she was in touch and for whom she had been drafting many of the formal questions that he asked in the United Kingdom's Parliament.

The MoD sent her the file, and once they did it soon became clear that the MoD's "no defence significance" line on Rendlesham was seriously compromised, because for years the MoD had implied—and sometimes specifically stated—that Charles Halt's one-page report was the only document that the department had on the incident. In a sense, this was farcical. Did people really expect that the USAF would report the fact that a UFO had landed next to two of the most sensitive bases in the NATO military alliance without there being some follow-up? Once Bruni had obtained the file, though, the cat was out of the bag. The file contained not

1 document but 190. While many of these were simply copies of letters from members of the public who had written in about the case, together with numerous versions of the MoD's ubiquitous "no defence significance" reply, there were some more interesting papers, too. The documents released included the DIS assessment that the radiation levels recorded at the landing site had been "significantly higher" than the expected background levels, papers relating to radar, and the embarrassing spat over the removal by General Gabriel of evidence relating to the case.

There was to be a final twist in the story of the MoD's evolving attempts to keep the lid on the Rendlesham Forest incident. The United Kingdom's Freedom of Information Act came fully into force in 2005, and it soon became clear that in the informal "league tables" of subjects on which people were making requests UFOs was at or near the top within the MoD. In 2007 the MoD decided to proactively release its entire archive of UFO files, to save having to respond to hundreds of FOIA requests on a case-by-case basis. This process started in 2008 and came to its conclusion in June 2013 when the final batch of documents was released. It involved the release of over fifty thousand pages of documentation, and while Georgina Bruni had secured the release of the UFO project's Rendlesham Forest file back in 2001, UFO researchers were looking forward in particular to the release of DIS UFO files covering the late 1980/early 1981 period that had not yet been released but would clearly also include documents on the Rendlesham forest incident. In March 2011, when the time came for the relevant date period to be released, the UFO community waited expectantly.

When the files were released, there was an embarrassing revelation. Out of all the UFO files, DIS files covering the period when the Rendlesham Forest incident took place had apparently been shredded. The ones covering the period immediately before the incident were there, as were the ones for the period immediately after. But the critical files had been destroyed. This was totally counter to policy, because as long ago as 1967 a decision had been made to keep *all* UFO files permanently, in view of their historical significance and public interest. To add insult to injury, the authorization slips that would have revealed when the files were destroyed, by whom, and on whose authority had also been destroyed. The trail, it seemed, had gone cold.

14. PROJECT CONDIGN

We have firmly established that the MoD's policy in relation to UFOs was to downplay both the phenomenon itself and the extent of the department's interest and direct involvement. This manifested itself in a number of ways, most usually by sending out standard letters to people who asked about UFOs or reported sightings, implying that the MoD had very little interest in the subject, which was regarded as being of "no defence significance/interest." Similar lines were trotted out to the media when the subject came up, and if pressed the department would try to ensure that its interest was inextricably linked with the question of foreign (generally Soviet—and later Russian) military aircraft probing at the United Kingdom's defenses. The latter was a very real scenario, so it was very easy to "spin" the MoD's involvement in this way.

If anybody thinks this is just semantics, this chapter deals with a situation that exposes the true reality of the situation. This is the story of Project Condign, a highly classified intelligence study into the UFO mystery carried out at the same time that the MoD was saying the subject was of no defense interest. The study specifically considered the Rendlesham Forest incident. All this can be demonstrated clearly by reference to documents declassified by the MoD and released under the Freedom of Information Act. In addition, I was personally involved in the work that led to this study being carried out, and indeed many of the documents released are ones that I wrote.

Project Condign had its roots in discussions that I had in 1993 with my opposite number in the DIS. While UFO sightings had been investigated since 1953, little trend analysis had been undertaken. To compound the problem, much of the information that we had on sightings was recorded in paper files and had not yet been entered into a computer. As a consequence, it was impossible to interrogate a single database and get answers to simple questions such as "Are there times of the year when UFOs are more commonly seen than others?," "What is the geographical distribution of UFO sightings in the UK and how does this change when allowing for population density?," "Have there been changes in the socio-economic status of UFO witnesses over the last ten years?," "Is there a correlation between UFO sightings and the theatrical release dates of sci-fi movies?," or any number of other such questions that we might usefully ask. The only way to get such answers was the very labor-intensive solution of assigning administrative staff to the task, which was not always possible, given the constant shortage of support staff. With around twelve thousand UFO sighting reports in the files we had a vast amount of data—far more than the US government had ever amassed during Project Blue Book—but no way of exploiting it properly.

Unlike many of my predecessors on the UFO project and most of my successors, I forged extremely close links with the DIS, possibly because I had a high security clearance, obtained as a consequence of having worked in the Air Force Operations Room in the Joint Operations Center, during the Persian Gulf War. My DIS opposite number seemed genuinely interested in the UFO phenomenon and shared my discomfort that we weren't doing enough trend analysis. So it was that on June 1, 1993, he wrote me an internal minute: "You may be interested to hear that at long last I have had some funds allocated for serious UFO research . . . needless to say we do not want this broadcast and it is for your information only."

I replied on June 3 and my response included the following: "I was pleased to hear about the funds you have secured, and stand ready to assist with any of the projects you are planning."

There then followed an internal DIS meeting at which it was decided that approval was required from "customers." The two customers identified were the head of Sec (AS)—my head of division—and the Director of Air Defence. A draft note was prepared that contained the following quote: "Some recent events, and a cursory examination of the files, indicate that the topic may be worthy of a short study." One of the "recent events" was a wave of sightings that had occurred in the United Kingdom on March 30/31, 1993, into which I had led the official investigation.

In the event, the draft letter seeking approval from the head of Sec (AS) and the Director of Air Defence wasn't sent. What happened was that a much more detailed briefing note was sent to Sec (AS) 2—the Deputy Director who was my second reporting officer. The reason for this was that approval was needed at either Director or Deputy Director level, so I wasn't able to approve the study myself. However, both my Director and Deputy Director were highly skeptical about UFOs and were uneasy that the UFO project was embedded in their division at all. How could we possibly convince them we should be doing *more*, not less, on the subject?

With this problem in mind, I had extensive discussions with my DIS opposite number and we concocted a strategy for "selling" the initiative to my skeptical bosses and getting "buy-in" and the green light. To do this, a little deviousness was required. First of all, my Deputy Director was a statistician, brought into the managerial mainstream as part of an MoD personnel department experiment to see if "deep specialists" could operate outside their comfort zone in wider policy-making and managerial roles. I therefore recommended that where possible, the study should be presented as a statistical analysis!

Language was a key part of this "selling job" and my DIS colleague and I agreed to drop the term "UFO" in view of the associated baggage, replacing it with the more scientific-sounding term "UAP" (Unidentified Aerial Phenomenon). It became a point of pride—and some humor—to write entire documents on the subject without the term "UFO" appearing even once. This was to have unanticipated consequences years later, when responding to Freedom of Information Act requests, and I'm sure that various UFO-related documents have not been released as planned simply because they use the term "UAP" and not "UFO."

With all this groundwork laid, the meeting was set up. Like many people with little experience of intelligence matters, my Deputy Director was a little overawed by the spooks (as Brits call intelligence operatives) and doubtless conjured up some James Bond fantasy. My DIS colleague played things masterfully, often speaking in little more than a whisper and leaning forward conspiratorially, as if he were entrusting my boss with the United Kingdom's nuclear launch codes. There was mention of Russian research in this area, a couple of uses of the term "UAP," and lots of talk about statistics. "Perhaps, if it's not too much trouble," the DIS briefer ventured, "we could call upon *your* expertise in this area—if we come up against a particularly tricky problem. . . ." My boss was almost salivating. At the end of the meeting, my boss was firmly "onboard." All that was needed to "clinch the sale" was written approval.

The written proposal was dated October 18, 1993, and was classified Secret UK Eyes A, only one level below the highest classification, Top Secret. The document set out a compelling case for undertaking the study, but there was to be an additional twist. To distance the MoD even further from this work, the detailed analysis was to be undertaken not by the DIS but by a defense contractor. At first, this might sound bizarre, but it should be borne in mind that on retirement senior military officers and defense civilians often step seamlessly from government service into the employ of large defense corporations, usually at board level. There's movement at lower levels, too, and the bottom line is that certain defense companies have a small group of individuals whose security clearances are still active and are generally higher than most of their ex-colleagues in the military, DoD, MoD, or wherever. If that isn't enough to set alarm bells ringing with conspiracy theorists, I should mention another benefit that arose from passing the work to a defense contractor: private companies are exempt from the Freedom of Information Act. This fit perfectly with the usual policy of denying any great interest in UFOs. The department could continue to make statements about "examining" sightings to see if there was evidence of a threat. Even the word "investigating" was frowned upon as being too strong a word, so we used to say that we "examined" or "logged" sightings. Quite how we were supposed to determine whether or not sightings were of "defense significance" without actually investigating them was never made clear! Passing the whole study off to a defense contractor would also enable us to make statements such as "The MoD has not undertaken any intelligence studies on UFOs" while desperately hoping that nobody would think to ask if we'd *commissioned* any! If that sounds like sharp practice, it should be remembered that a golden rule in government is to answer the question that's been asked, not the question that somebody *meant* to ask or *should* have asked.

Using defense contractors to undertake particularly sensitive classified tasks is not without precedent when it comes to esoteric subjects that the MoD has dabbled with. Consider this statement, placed in the Freedom of Information section of the MoD's Web site: "A study was undertaken in 2001–2002 to investigate theories about capabilities to gather information remotely about what people may be seeing and to determine the potential value, if any, of such theories to Defence."

At first, this might sound as if it relates to some new surveillance technique. In fact, this study relates to another subject on which I had discussions with DIS colleagues: recruiting psychics! The MoD had been extremely reluctant to release the study at all but could not avoid its legal responsibilities under the Freedom of Information Act. So when the material was published, it was done under cover of a bland note that deliberately avoided using words such as "psychic," "clairvoyant," and "ESP" to avoid being picked up by Internet searches using such keywords and to lessen the chances that anyone in the media would notice a document classified Secret UK Eyes Only, describing how, in the immediate aftermath of 9/11, the MoD had attempted to lure psychics into a secret government program to locate "targets of interest."

Returning to the UFO study, in view of the extreme sensitivities, even the normal procedures for placing a defense contract were to be bypassed: "I believe that opening a new contract especially for this study and using competitive tendering would potentially expose the study to too wide an audience. . . ." The plan was to amend an existing contract, so as to avoid creating any paper trail. All this, of course, was planned well before the United Kingdom's Freedom of Information Act, and there was no intention that this study would ever see the light of day.

Having gotten my skeptical Deputy Director to agree to something that he would not normally have supported, I was eager that we sign off on the request as soon as possible and with the most low-key of responses. I sent my boss a note downplaying what I knew was involved in the study and stating that if the DIS was going to pay for the study, it was essentially their business. This wasn't strictly true, as we had the policy lead in terms of UFOs. However, highlighting the fact that something isn't going to come out of your budget always goes down well:

> *The attached note from DI55c seems fine. AD DI55c came over to discuss this some time ago, and I advised him that I saw no difficulties, although I asked him to write to us setting out their proposals; their money, their business.*
>
> *As our policy already commits us to looking at all reports to ensure there is no threat to the defence of the UK, we are not in any way going outside our remit, as this study is no more than a review of data.*
>
> *As you can see, the term "UFO" is being dropped, as it is rather a loaded term.*
>
> *Coincidentally, this week we have received a new report of a huge triangular object flying over the UK mainland.*
>
> *I have attached a short draft reply to DI55.*

With the preceding in mind, I not only briefed my Deputy Director on the DIS proposal but also made things easy for him by drafting a reply for him to send in response to the proposal. All he had to do was sign it, which he did: "I can confirm that we are content with what has been proposed . . . I would be grateful if you would keep Sec(AS)2a [i.e., me] involved in this process." The final part of the response was drafted with the intention that my Deputy Director would play no further part in the study. It was far better to handle such detailed work on an SME (Subject Matter Expert) to SME basis.

A series of delays caused by financial constraints and the pressure of other competing DIS priorities derailed the original plan. Correspondence dragged on periodically, but it was looking less and less likely that the study would ever take place. Despite the fact that some staff within the DIS felt "we have a remit that we have never met" (i.e., undertaking a proper intelligence study of the UFO phenomenon that would allow a better and more-informed assessment of whether UFOs were of defense significance), other DIS personnel were less keen. In a letter dated October 25, 1995, a senior DIS officer said that "spending money on such an esoteric subject in a continuing climate of constraint was not good politically."

However, by the end of 1996 the study was resurrected and, for the first time, the phrase "Project Condign" was used. The name was a randomly generated code name. The United Kingdom tends to do this, as opposed to setting out their intention in operation names in the way the US government does. Hence, the United Kingdom's version of Operation Desert Storm was Operation Granby, while the Falklands War was Operation Corporate. Conspiracy theorists didn't see it that way and, when the report was declassified, thought that the similarity with the "Condon Report" was no coincidence. The Condon Committee had reviewed Project Blue Book in the late sixties and the skeptical final report recommended that the USAF cease all research and investigations into the UFO phenomenon, which they duly did. This was exactly the outcome that the USAF had hoped to get, and testimony from some of the people involved in the Condon Committee suggests that the skeptical conclusions and recommendation were a foregone conclusion—exactly as conspiracy theorists allege.

By early 1997 another problem emerged: Sec(AS) began to get cold feet. "I have some concerns about what is planned" stated a letter dated January 27, 1997, before setting out worries that the DIS study would be incompatible with the "we don't investigate UFOs" line that Sec(AS) was aggressively trying to push with Parliament, the media, and the public. The DIS, however, had a concern of its own. The DIS concern was Sec(AS)! Despite the classification and sensitivity of this study, which was carried out on the usual "need to know" basis, Sec(AS) had copied the correspondence more widely. They had even included the name of the defense contractor undertaking the work, who delivered a stinging rebuke to Sec(AS).

I hasten to add that by that time I was no longer working in Sec(AS), having been promoted in 1994 and posted to another directorate. Had I still been in Sec(AS) I like to think I could have smoothed over these differences. I certainly would have advised my bosses against putting DIS documents on a wider distribution without their prior consent. The tone of the correspondence at this time clearly shows some tension between Sec(AS) and the DIS. And that's when the really interesting thing happens: Sec(AS) drops out of the picture altogether. The reason is never directly explained, but an internal DIS e-mail dated December 17, 1999, suggests that Sec(AS)'s wider circulation of DIS papers was the straw that broke the camel's back. Project Condign had been completed and the DIS e-mail was discussing circulation of the final report. The key quote is as follows: "No positive purpose would be served in sending the report to Sec(AS) . . . we recommend a letter to Sec(AS) . . . identifying that we have completed our declared review, outlining the conclusions drawn . . . in view of the 'leakiness' of Sec(AS) we would advocate only releasing the report to them on request, in order to discourage further discussion." Needless to say, for one part of the MoD to describe another part using the word "leakiness" is little short of sensational.

However, by 2000 things had changed again and while the letter summarizing Project Condign was sent to a handful of DIS and RAF personnel, it wasn't copied to Sec(AS) at all. MoD personnel tend to move posts (either on level transfer or promotion) every three or four years, so by 2000 none of those involved in the original discussions concerning Project Condign were in Sec(AS). This led to the extraordinary and farcical situation where a highly classified UFO study had been produced and was sent to a number of MoD divisions but not to the UFO project! The very division that had policy responsibility for the subject was completely excluded. The only (probably unintended) benefit of this was that it if Sec(AS) made statements such as "We are not aware of any classified studies into the UFO phenomenon" such statements would be true!

While the DIS had done their best to hide the existence of Project Condign not just from the public but even from the MoD's UFO project, they were ultimately unsuccessful. When the United Kingdom's Freedom of Information Act came fully into force on January 1, 2005, the MoD soon found itself receiving more requests on UFOs than on almost any other subject. Unfortunately (for the MoD), some apparently innocuous UFO-related documents that had been released made passing reference to a "policy review" into the UFO phenomenon that had been undertaken in around 1997. While this throwaway line was fairly cryptic and the date was wrong, it was enough to attract the attention of various ufologists, who submitted further FOI requests relating to this "policy review." One can only imagine the embarrassment and humiliation in the UFO project when they discovered that contrary to their belief that no such review had been carried out (because they would have written it—or at least seen it), such a review had been undertaken, but without their being involved or even informed.

The MoD duly released a redacted copy of the final report and published the document on its Web site on May 15, 2006. The statement on the MoD's Web site read as follows:

> During a policy review in 1996 into the handling of Unidentified Aerial Phenomena sighting reports received by the Ministry of Defence, a study was undertaken to determine the potential value, if any, of such reports to Defence Intelligence. Consistent with Ministry of Defence policy, the available data was studied principally to ascertain whether there is any evidence of a threat to the UK, and secondly, should the opportunity arise, to identify any potential military technologies of interest.
>
> The Ministry of Defence has released this report in response to a Freedom of Information request and we are pleased to now make it available to a wider audience via the MoD Freedom of Information Publication Scheme. Where indicated information is withheld in accordance with Section 26 (Defence), Section 27 (International Relations) and Section 40 (Personal Information) of the Freedom of Information Act 2000.

There was much media speculation about the anonymous author of the report, with many people suggesting that I had written it. I had not, though the identity of the author is known to me, as is the defense company involved.

Project Condign's final report was a bizarre document, to say the very least. As with many highly classified documents, the most intriguing material was not necessarily obvious, perhaps by accident or perhaps by design; there's an old saying that the best place to hide a book is in a library—or, in a version that the Rendlesham Forest incident witnesses might better appreciate, the best place to hide a tree is in a forest.

The report ran to 465 pages and only eleven copies were made at the time. It was an odd mixture: a comprehensive drawing together of some existing research on the topic coupled with some exotic new theories: "That UAP exist is indisputable. Credited with the ability to hover, land, take off, accelerate to exceptional velocities and vanish, they can reportedly alter their direction of flight suddenly and clearly can exhibit aerodynamic characteristics well beyond those of any known aircraft or missile—either manned or unmanned."

So stated the Executive Summary, bullishly, before going on to say, more cautiously, that no evidence has been found to suggest UAP are "hostile or under any type of control." But by its own admission, the study had not turned up a definitive explanation for the phenomenon: "Although the study cannot offer the certainty of explanation of all UAP phenomena . . . ," it stated, leaving the door open.

One of the most contentious aspects of the study related to what the report referred to as "plasma related fields." Electrically charged atmospheric plasmas were credited with having given rise to some of the reports of vast triangular craft, while the interaction of such plasma fields with the temporal lobes in the brain was cited as another reason why people might feel they were having strange experiences. The problem with this statement is that there's no scientific consensus here, and as a good rule of thumb one shouldn't try to explain one unknown phenomenon by citing evidence of another. In other words, you can't explain one mystery with another one.

Nonetheless, once the issue was raised, the potential military applications (including weaponization) of this could not be ignored. The report stated: "There is evidence . . . that scientists in the former Soviet Union have taken a particular interest in 'UFO Phenomena.' They have identified the close connection with plasma technologies and are pursuing related techniques for potential military purposes. For example, very high power energy generation, RF Weapons, Impulse Radars, air vehicle drag and radar signature reduction or control, and possibly for radar reflecting decoys." On this point, the report recommended: "The relevance of plasma and magnetic fields to UAP was an unexpected feature of the study. It is recommended that further investigation should be . . . [undertaken] into the applicability of various characteristics in various novel military applications."

The report also dealt with air safety issues, though incredibly when the report was released in 2006 the media failed to pick up on speculation that several fatal aircraft accidents might be attributable to pilots taking violent evasive action to avoid UFOs. Project Condign's final report made two recommendations in relation to pilots encountering UFOs: The first said: "No attempt should be made to out-maneuver a UAP during interception." Another recommendation stated: "At higher altitudes, although

UAP appear to be benign to civil air-traffic, pilots should be advised not to maneuver, other than to place the object astern, if possible."

Most relevant to us, however, is what the Condign Report had to say about the Rendlesham Forest incident. This is specifically addressed in volume 2 of the report, in annex F of Working Paper 1. Under the heading "Non-ionizing EM Effects on Humans" and "EM Field from a Plasma" the report states: "The well-reported Rendlesham Forest/Bentwaters event is an example where it might be postulated that several observers were probably exposed to UAP radiation for longer than normal UAP sighting periods. There may be other cases which remain unreported. It is clear that the recipients of these effects are not aware that their behavior/perception of what they are observing is being modified."

In describing the effects of exposure to such radiation, the report goes on to say: "An important fact is that the reported effect of (presumed) UAP radiation on humans is that it is quick acting and remembered—although, curiously, following the event there is little or no recall of events as a continuum. In short, the witness often reports apparent 'gaps' or 'lost time'—often not accounting for up to several hours. It is described as if the exposure causes a temporary memory erasure."

The Executive Summary put further flesh on the bones by saying this:

The close proximity of plasma related fields can adversely affect a vehicle or person. For this to occur, the UAP must be encountered at very close ranges. A probable modulated magnetic, electric or electromagnetic (or even unknown) field appears to emanate from some of the buoyant charged masses. Local fields of this type (probably either an electromagnetic near-field, or a direct magnetic field), have been medically proven to cause responses in the temporal lobes of the human brain. These result in the observer sustaining (and later describing and retaining) his or her own vivid, but mainly incorrect, description of what is experienced. Some observers are likely to be more susceptible to these fields than are others, and may suffer extended memory retention and repeat experiences. This is suggested to be a key factor in influencing the more extreme reports found in the media and are clearly believed by the "victims."

So, ironically, whatever one thinks of the theory, the UK MoD at least offers a view on what happened to the witnesses at Rendlesham, when those same witnesses can't even get the US government to comment on UFOs, let alone offer some potential explanation to what happened to their own military personnel in Rendlesham Forest. Penniston certainly draws some comfort from the study: "Project Condign, in some respects, was a validation of the existence of the phenomenon."

Burroughs is even more forthright on this point:

> *Project Condign has not been looked at very closely by most people because of the way it was written and because of the fact that most people don't have the background to understand what it's trying to tell you. Only after you take a look at the documents used to write the report do you begin to understand what the Governments of the world are working on and why they want this to remain classified. If you take the time to read the documents you will see not only that Rendlesham is mentioned but that there are numerous references in these documents talking about what we encountered in the forest—a phenomenon that, if harnessed, could be used as a weapon. What's also interesting is that many of the documents used in this report remain classified.*

Burroughs elaborates on this point: "If you look at the MoD papers it's very clear there is a race by many countries to get their hands on the technology that we encountered over those three nights. And yet, so far as most people in the world are concerned, they just want to know if we're alone or not." In restating this point, Burroughs sets out his view on just how high the stakes are: "It's very clear there is a race, around the world, to have the upper hand in technology which would then give that government an upper hand not only on its own people, but on the entire world."

Simply put, technology acquisition lies at the heart of government interest in UFOs, and explains UFO secrecy. And again, as articulated by Burroughs, comes the intriguing possibility that the extraterrestrial hypothesis could simply be a cover story masking an even bigger secret: "So far as the story of aliens is concerned, that was just a cover for what really happened to us."

There was to be one final surprise. Officially, as we've seen, the US government has not investigated UFOs since Project Blue Book was terminated in 1969. Yet in 1993, while making the case for the study to be carried out, the DIS stated: "I am aware, from intelligence sources, that XXXXX believes that such phenomena exist and has a small team studying them. I am also aware that an informal group exists in the XXXXXXXXXXX community and it is possible that this reflects a more formal organization." Investigative journalist Leslie Kean conducted a detailed analysis of this document (formerly classified Secret UK Eyes A and partially declassified under the Freedom of Information Act) and concluded that the first redaction is "Russia" and the second redaction is "US intelligence."

I clearly recall the document but in view of my secrecy oath can neither confirm nor deny Kean's analysis. However, if she's right, it would mean that conspiracy theorists are, for once, correct: someone in the United States *is* still carrying out research and investigation into the UFO phenomenon, despite the official denials.

15. BEYOND RENDLESHAM

The MoD's Project Condign certainly made for bizarre reading, with its theories about atmospheric plasmas, non-ionizing electromagnetic (or "unknown") fields, and the effects of all this on the temporal lobes in the human brain. The suggestion that the Rendlesham Forest witnesses "were probably exposed to UAP radiation for longer than normal UAP sighting periods" certainly has frightening implications for John Burroughs and Jim Penniston. But for an organization fond of labeling UFOs as being of "no defence significance" to recommend that "no attempt should be made to out-maneuver a UAP during interception" seemed odd. It was a classic example of the Orwellian "doublethink" that the MoD occasionally had to practice on this subject: we had one (dismissive) view on UFOs for our dealings with Parliament, the media, and the public but another, more complex position for ourselves.

Project Condign's "No attempt should be made to out-maneuver a UAP during interception" recommendation was an example of this, and indeed the final report stated that "the flight safety aspects of the findings should be made available to the appropriate RAF Air Defence and other military and civil authorities which operate aircraft, particularly those operating fast and at low altitude." Criminally, given the report's reference to fatalities, there is no evidence to suggest that the UK Civil Aviation Authority or the US Federal Aviation Administration (FAA) has been briefed on any of this, despite the recommendation.

Why is any of this important? The answer lies in another finding from Project Condign, which again, incredibly, was missed by the media (proving that the best place to hide a book really is in a library): "Attempts by other nations to intercept the unexplained objects, which can clearly change position faster than an aircraft, have reportedly already caused fatalities." Perhaps it's only when one replaces the term "unexplained objects" with the phrase the MoD was so desperate to drop—"UFOs"—that one can see just how explosive a statement that is. Never has the phrase "no defence significance" seemed so inappropriate.

Before looking at some other UFO cases that will put the Rendlesham Forest incident into context by showing the sorts of UFO incidents that governments take seriously (whether or not they say so), it's worth looking at who, within government, actually does the investigating. At first, this might seem like an odd point to highlight, but it gets to the heart of the issue because it's one of those fundamental questions that hardly ever get asked. Let me put it another way. If the US government really did get out of the UFO "game" after Project Blue Book was terminated but was going to look again at the phenomenon, which agency should be put in charge? The possibilities include NASA, the FAA, and the Office of Science and Technology Policy. But as we have seen, Project Blue Book was embedded in the USAF. Indeed, with the exception of the French government (whose UFO project, known as GEIPAN, is embedded in the French National Space Agency), virtually every country that has ever investigated UFOs has charged their military (usually the Air Force) or their DoD with this work. So the question about who investigates UFOs is an important one and the answer gives us an insight into how UFOs are really seen: as a defense issue. So with that in mind, let's dip into some UFO case files from America, Britain, and elsewhere. Because while most of the cases in these real-life X-Files are mundane sightings that turned out to be misidentifications of things like aircraft lights, weather balloons, or satellites, others are decidedly more mysterious—and, on occasion, sinister.

A FATAL ENCOUNTER

On January 7, 1948, Kentucky Air National Guard pilot Thomas Mantell was one of a number of pilots who chased a UFO that had been seen from various locations in Kentucky. While the other pilots broke off their pursuit, Mantell carried on but is assumed to have blacked out due to the lack of oxygen. His plane crashed and he was killed. The US military later said he had mistakenly been chasing the planet Venus.

SHOOT DOWN THE UFO

On or around May 20, 1957 (the precise date is the subject of some debate), Milton Torres, a USAF pilot based in the United Kingdom as part of a US/UK exchange program, was scrambled and ordered to intercept a UFO that was being tracked on military radar. As he closed on the object in his F-86D Sabre fighter jet, he was ordered to open fire on it. He recalls that he was at a height of around thirty-two thousand feet and that while he never saw the UFO, his aircraft's onboard radar tracked the object, which appeared to be the size of an aircraft carrier. He came within seconds of firing off a full salvo of twenty-four "Mighty Mouse" rockets, but at that point the UFO accelerated from a virtual hover to a speed of around Mach 10, causing him to break off his attack run.

Torres stated that he was subsequently warned to stay silent about the incident and told that if he failed to comply with this order he would be grounded. He does not know the identity of the person who debriefed him and made this threat but now believes the individual concerned was from the NSA. For pilots, "losing their wings" is virtually the ultimate threat. Torres stayed silent until long after he retired, when he mentioned the incident at a reunion. Torres then wrote to the MoD, in 1988, sending an account of the event and asking whether it was now all right to discuss the incident and asking whether any definitive explanation for the encounter had ever been found and, if so, whether it was available. No original documents relating to the case were discovered—the MoD has long maintained that most pre-1967 UFO files were destroyed many years ago.

The story only emerged in 2008, when the MoD file containing the letter from Torres was released, as part of the wider program to declassify and release all the MoD's UFO files. I had left the MoD by then and subsequently gave some media interviews about this incident. In 2009 I gave a talk in Washington, D.C., about some of my government work on the UFO phenomenon. Torres, by then seventy-seven years old, shared the stage with me and told his story with quiet dignity. He explained that he had stayed silent for decades, as ordered, out of loyalty. He said that he had not even told his father, who he felt sure would have loved to hear the story but had died before the MoD released the file. At that point Torres broke down. It was a sad indication of the stresses of keeping UFO secrets for so long.

ANOTHER ATTEMPTED SHOOT-DOWN

On September 18, 1976, Parviz Jafari—who later retired as a general in the Iranian Air Force—attempted to shoot down a UFO with an air-to-air missile. The UFO had been seen over Tehran and an air force commander ordered a jet to be scrambled. As Jafari approached the UFO in his Phantom F-4 jet he locked on to it with his airborne radar. As he approached, he saw four smaller objects detach from the main craft. One of them came rapidly toward his aircraft. Believing he was under attack, Jafari attempted to launch a heat-seeking Sidewinder missile, but at that instant his missile control panel went dead.

On his return to base, Jafari was debriefed by a number of people, including an American colonel—this was Iran under the Shah, when the countries were allies. Years later Jafari saw a DIA document on his encounter, released under the US Freedom of Information Act. It stated: "This case is a classic that meets all necessary conditions for a legitimate study of the UFO phenomenon." Again, it's noteworthy that this was six years *after* the termination of Project Blue Book and thus a time when the US government was telling the media and the public that it had no interest in UFOs and was no longer investigating sightings. If that was true, why was a US colonel quizzing Jafari about a UFO sighting and why was the DIA writing a classified report on the incident?

The UFO report had come to the attention of the US government via the Military Assistance Advisory Group in Tehran. This organization was headed by the USAF general Richard Secord. Though outside the scope of this book, it's interesting to note that Secord was allegedly part of an informal group of defense and intelligence community officials dubbed the

Enterprise. Other members of the Enterprise were said to include Edwin Wilson, Thomas Clines, Theodore Shackley, and Erich von Marbod. The existence of the Enterprise came to light in the course of investigations into the Iran Contra affair. I mention the Enterprise because of the Secord/Jafari connection and because it's exactly the *sort* of body (though probably not the specific one) that the UFO issue might have been handed off to, in a way that would keep it far enough away from government to be "off-the-books" but close enough to control.

MISSING PRESUMED DEAD

On October 21, 1978, Frederick Valentich, a civilian pilot, took off from Melbourne, Australia. He had been in the air for around forty-five minutes when he radioed the Melbourne Flight Service Unit to report that he was being buzzed by a strange, unidentified aircraft. He said it was large, bright, and metallic. His last radio message to the air traffic controller was as follows: "My intentions are to go to King Island—er, Melbourne, that strange aircraft is hovering on top of me again—it is hovering and it's not an aircraft." There was no further contact, and despite an extensive search-and-rescue operation, neither Valentich nor his aircraft was ever seen again.

TARGET ENGAGED

On April 11, 1980, Oscar Alfonso Santa Maria Huertas, a pilot with the Peruvian Air Force, was scrambled to intercept a spherical UFO hovering in restricted military airspace. His unit commander's orders were clear: "Shoot down the UFO." As Huertas closed on the object in his Sukhoi-22 fighter, he strafed it with his 30mm cannon, firing sixty-four shells at the UFO. He saw some of his projectiles hit the craft, but they had no effect. "The projectiles didn't bounce off," he said. "Probably, they were absorbed." He vectored his aircraft for another attack, but on this occasion the UFO took evasive action and he was not able to catch up with it again.

THE CASH-LANDRUM ENCOUNTER

On December 29, 1980, three witnesses (Betty Cash, Vickie Landrum, and Vickie's seven-year-old grandson, Colby Landrum) in a car encountered a huge flaming diamond-shaped UFO in Dayton, Texas. The UFO was at treetop height and the heat was intense. The two adults got out of the car for a closer look, but Vickie soon returned to comfort the hysterical Colby. Betty stayed outside, as if mesmerized by the light, and by the time she returned she said the door handle was so hot, she had to use her coat to protect her hand. A large number of military helicopters were seen flying alongside the UFO, as if they were escorting it. Subsequently, all three witnesses became ill, with nausea and vomiting. The most badly affected witness (Betty) was hospitalized for twelve days with symptoms that doctors believed were consistent with radiation sickness.

Eventually, Cash and Landrum launched a legal action, claiming $20 million from the US government. The action was dismissed after statements from various branches of the US military were produced denying that they owned or operated any craft along the lines of the one described. Assuming these denials were true, is it possible that such craft are operated not by the US government or military but by a private corporation or by a quasi-governmental organization such as the Enterprise? Just as highly-classified MoD studies such as Project Condign and the secret project to recruit psychics were moved into the private sector, not least to take it outside the scope of the Freedom of Information Act, perhaps the same is true of certain secret prototype aircraft and drones. UFO believers, of course, think the craft that irradiated the Cash-Landrum party came from considerable farther afield than, say, the Lockheed Martin Skunk Works.

Betty Cash died on December 29, 1998, eighteen years to the very day since the incident occurred. Vickie Landrum died on September 12, 2007.

Sharp-eyed readers will have spotted that the date of this encounter—December 29, 1980—was just a few hours after the events at Rendlesham Forest had come to their climax when Charles Halt and others encountered a UFO that fired light beams down at them and, later, at the Bentwaters WSA. John Burroughs and Jim Penniston are intrigued by this coincidence (if indeed that's what it is) and concerned at the suggestion that the UFO in the Cash-Landrum encounter exposed the witnesses to radiation that may, ultimately, have proven fatal. This is another reason why, so many years after their encounter, Burroughs and Penniston are still lobbying for answers from the US government. Penniston puts it this way: "It is interesting the two events happened within such a short time of each other. I find the Rendlesham Forest incident more than enough for me to handle, let alone others."

TRANS-EN-PROVENCE

On January 8, 1981, a farmer working in a field near the town of Trans-en-Provence in southern France heard a strange whistling sound and saw a saucer-shaped UFO land nearby. The craft quickly took off and left what seemed to be burn marks on the ground. The investigation concluded that the object had weighed around five tons, had heated the soil to a temperature of around six hundred degrees Celsius, and had caused bizarre changes to the nearby vegetation, including loss of chlorophyll.

THE BELGIAN TRIANGLE

On the night of March 30/31, 1990, a wave of UFO sightings in Belgium that had lasted for several months reached its climax when an unidentified object was picked up on military radar. The Belgian Air Force scrambled two F-16 fighter jets to intercept the craft, and the pilots achieved lock-ons with their onboard radars. Witnesses on the ground said the UFO was a vast triangular craft, the size of a football field. The UFO evaded them as a bizarre game of cat and mouse was played out in the skies over Belgium. The Belgian Air Force launched an official inquiry, but no explanation was ever found.

THE CALVINE ENCOUNTER

On August 4, 1990, two members of the public were out walking in the vicinity of Calvine, near Pitlochry, in Scotland when they sighted a massive diamond-shaped metallic UFO. The UFO was virtually stationary and hovered silently for what the witnesses believed was several minutes before accelerating away vertically at massive speed. During the sighting, a military aircraft, believed to be a Harrier, was seen, but it wasn't clear if the military aircraft was escorting the craft or attempting to intercept it or whether the pilot was ever aware of it at all.

A number of color photographs were taken and passed to a Scottish newspaper, whose staff contacted the MoD, presumably because they were seeking a comment for a story. It's not clear what happened next—the declassified files are unclear on this point, and because I didn't join the MoD's UFO project until 1991 this investigation was handled by my predecessor. It seems that, somehow, the MoD managed to persuade the reporter to part with not just the photos but also the negatives.

The photos were then sent to the DIS, who then sent them on to imagery analysts at JARIC (the Joint Air Reconnaissance Intelligence Centre), despite the fact that at the time the MoD had not yet publicly acknowledged that there was any intelligence interest in UFOs at all. I asked my DIS opposite number about the image. I was told that the official assessment was that the photos were real and the craft had a diameter of around eighty feet. Despite this sensational conclusion, MoD documents drawn up when it was feared that the media might run a story about the photos show that if any journalists contacted the MoD Press Office the line to take on this was to be that "no definite conclusion had been reached regarding the large diamond-shaped object."

I first came across this story in 1991, when I joined the UFO project. A poster-sized enlargement of the best photo was prominently displayed on the office wall. I worked in a four-person office and my predecessor had put it up. It was one of the few visible UFO-related items on display; most such material was locked away. The office dealt with some other issues, too; most of us had been seconded into the Air Force Operations Room during the Persian Gulf War. One of my other jobs was to read draft book manuscripts that included any RAF aspects of the war to ensure nothing was published that was classified, detrimental to the RAF or the MoD, or politically embarrassing. Sometimes people would come to our office to discuss non-UFO business, and some of these people were not aware that the UFO project was embedded in the section. You would have this surreal moment when people would stop mid-sentence, stare at the Calvine UFO photo, and say something like, "What the hell's that?" This was not the archetypal distant, blurred UFO photo. This was "up close and personal, reach out and you can touch" it stuff. "I don't know what it is, but it's not one of ours," was our stock answer to the inevitable question. At a DIS briefing on UFOs that I took my boss to, the briefer took out his own copy of the photo. "Take this, for example," he began. "It isn't American and it isn't Russian." As he said "American" he pointed to one side and as he said "Russian" he pointed to the other. "So that only leaves . . ." His voice trailed off and he did not complete the sentence. But his finger was pointing directly upwards.

At some point in 1994 my head of division removed the photograph and locked it in his office safe, telling me that he believed the craft photographed was probably a secret, prototype aircraft or drone. What happened next? The suspicion is that someone shredded the photo, but whatever the truth of the matter, it was never seen again. As we have seen, the same thing had happened to the DIS files on the Rendlesham Forest UFO incident, the Highpoint Prison and Hollesley Bay Youth Correction Centre governors' journals, and the HMS *Manchester* ship's log. This was some years before the United Kingdom got its Freedom of Information Act. At the time, shredding the photo—if that's what happened—would probably have been a legitimate (albeit unfortunate) action. If such an action happened post-FOIA and was a deliberate attempt to circumnavigate the Act, it would have been illegal.

When the story broke in March 2009, as part of the media coverage of the program to declassify and release the archive of MoD UFO files, I gave various interviews about this, giving my recollection of events. Despite this media coverage and associated public appeals, the witnesses have never come forward. Neither has anyone at the newspaper concerned (or any other Scottish newspaper) come forward to say that they worked on this story back in 1990. Understandably, this has generated a few conspiracy theories. I suspect that in their desperation to acquire the photos and negatives (and maybe kill the media story) DIS staff somehow tricked the journalist into handing over all the material and never gave it back. If the journalist had not briefed his editor, he may have stayed silent out of embarrassment. Similarly, maybe the witnesses were told that it would be better if they did not discuss what they had seen (something that MoD staff sometimes suggested when engaging with UFO witnesses), took this to be a threat, and complied.

The MoD files that contain documents relating to this case have been released and are available at the United Kingdom's National Archives. The MoD says that despite an extensive search, no trace has been found of the images, aside from one poor-quality photocopy of a line drawing that was done as part of the original MoD investigation.

OVERTAKEN BY A UFO

On November 5, 1990, a number of RAF Tornado fast jet aircraft flying over the North Sea were casually overtaken by a UFO. The pilot's report stated, in part: "UFO appeared in our right hand side . . . we were travelling at Mach point 8. It went into our 12 o'clock and accelerated away. Another 2 Tornados saw it."

NEAR MISS BETWEEN AIRCRAFT AND UFO

One of the most disturbing UFO cases with which I have been personally involved occurred on April 21, 1991. We were informed by the Civil Aviation Authority (CAA) that there had been a near collision between a commercial aircraft and an unknown object. The aircraft concerned was an Alitalia MD-80 with fifty-seven passengers onboard. It was at a height of around twenty-two thousand feet over Kent, near Lydd, when a brown cigar-shaped object passed so close to the aircraft that the pilot shouted, "Look out! Look out!" In the normal course of events, any near miss would be investigated by the CAA. However, most such incidents involve other aircraft, and as the crew was not able to identify the object, it was treated as a UFO incident and passed from the CAA to the MoD. We launched a full investigation and eliminated all the usual possibilities, including weather balloons, military aircraft, et cetera. We even checked to see whether we had accidentally fired off a missile of some sort. We drew a complete blank and the incident remains unexplained to this day.

THE COSFORD INCIDENT

Aside from the Rendlesham Forest incident, the United Kingdom's most famous and compelling UFO case is a series of sightings that took place over about six hours on March 30/31, 1993. These sightings are known collectively as the Cosford incident, and as this incident took place during my time on MoD's UFO project, I led the investigation.

The first sighting took place on March 30 at around 8:30 pm in the county of Somerset. This was followed by a sighting at 9:00 pm in the Quantock Hills. The witness was a police officer who, leading a group of scouts, had seen a craft that he described as looking "like two Concordes flying side by side and joined together." The reports came in thick and fast and it was soon clear that I had a major UFO event on my hands. One of the most interesting reports came from a member of the public in Rugely, Staffordshire, who reported a UFO that he estimated as being two hundred meters in diameter. He and other family members told me how they had chased the object in their car and gotten extremely close to it, believing it had landed in a nearby field. When they got there a few seconds later, there was nothing to be seen. Many of the descriptions related to a triangular craft or the lights perceived as being on the underside of such a craft.

Some of the reports received that night read like something from *The X-Files*. A couple walking home past a farmer's field in the West Country saw the UFO, which they thought was so low that it was going to land. When they got to the field, it was empty. However, all the farmer's cows were standing silently in the middle of the field, facing one another in a perfect circle.

The UFO was seen by a patrol of RAF Police based at RAF Cosford. Their official police report (classified "Police in Confidence") stated that the UFO passed over the base "at great velocity . . . at an altitude of approximately 1000 feet." The report described two white lights with a faint red glow at the rear, with no engine noise being heard. The report also contained details of a number of sighting reports from civilian witnesses. They had been made aware of these reports in the course of making inquiries with other military bases, civil airports, and local police.

Later on that night, the Meteorological Officer at a second air force base, RAF Shawbury, saw the UFO. He described to me how it had moved slowly across the countryside toward the base, at a speed of no more than 30 or 40 mph. He saw the UFO fire a pencil-thin beam of light (like a laser) at the ground and saw the light sweeping backward and forward across the fields beyond the perimeter fence, as if it was looking for something. He heard an unpleasant low-frequency humming sound coming from the craft and said he could *feel* this sound as well as hear it—as if he were standing next to a huge bass speaker. He estimated the size of the craft to be midway between a C-130 transport aircraft and a Boeing 747. Then he told me that the light beam had retracted in an "unnatural" way and that the craft had suddenly accelerated away to the horizon many times faster than a military jet fighter. The witness was an experienced RAF officer who had been in the military for eight years.

I launched a detailed investigation into these sightings, working closely with the RAF, colleagues in the DIS, and personnel at the Ballistic Missile Early Warning System at RAF Fylingdales. One of the first things that I did was order that radar tapes be impounded and sent to me at MoD Main Building in Whitehall. The film of the radar tapes was downloaded onto VHS videocassettes and arrived shortly thereafter. I watched it with the relevant RAF specialists, who told me that there were a few odd radar returns but that they were inconclusive. Later a more formal assessment of the radar data was made. Unfortunately, one of the radars was not working fully on the night in question (echoing what the UFO project had been told happened with some of the radar systems during the Rendlesham Forest incident), though we had enough data so that, taking into account certain other checks, I was able to build up a comprehensive picture of all aircraft and helicopter activity (civil and military) over the United Kingdom and eliminate this from the investigation.

The Ballistic Missile Early Warning System at RAF Fylingdales, with its powerful space-tracking radars, was an important part of my UFO investigation. Staff at Fylingdales quickly alerted me to the fact that on the night in question there had been a re-entry into the Earth's atmosphere of a Russian rocket carrying a communications satellite, Cosmos 2238. We postulated that this was a possible explanation for a cluster of UFO sightings that occurred at around 1:10 am on March 31. But the sightings had taken place over a period of six hours and the most spectacular events, detailed earlier, were at different times.

After my investigation had eliminated all possible conventional explanations, I prepared a briefing for my head of division. In the document, dated April 16, 1993, I wrote a conclusion deliberately designed to contradict our public line on UFOs: "It seems that an unidentified object of unknown origin was operating in the UK Air Defence Region without being detected on radar; this would appear to be of considerable defence significance, and I recommend that we investigate further, within MoD or with the US authorities."

My head of division was normally skeptical about the UFO phenomenon, but on this occasion he agreed with my conclusion. On April 22, 1993 he wrote to the Assistant Chief of the Air Staff (one of the United Kingdom's most senior military officers) and concluded: "In summary, there would seem to be some evidence on this occasion that an unidentified object (or objects) of unknown origin was operating over the UK."

There was an extraordinary twist in this tale. My head of division had somehow convinced himself that this UFO sighting was attributable to Aurora—the name given to a supposedly secret, prototype hypersonic US aircraft that was believed to be the intended replacement for the SR-71 Blackbird. Rumors about Aurora had been circulating for a few years in aviation magazines and UFO newsletters, and my head of division thought the existence of Aurora might explain the Cosford incident and, perhaps, some of the United Kingdom's other more intriguing UFO sightings. But there was a problem with this theory: we had raised this question with US authorities, but they had told us categorically that no such aircraft had been flown over the United Kingdom, and on the basis of these assurances UK defence ministers had told Parliament that no such aircraft had been flying over the United Kingdom. There would be a domestic political scandal and an international incident between the United States and the United Kingdom if these various assurances turned out to be false. Despite this, various colleagues in the UFO project and the DIS did not believe the United Kingdom's closest international ally and a declassified MoD document on the subject said that we "would not be surprised if it did exist." The concerns were again put to HQ USAF by staff at the British embassy in Washington. The Americans were incandescent with rage. A December

22, 1992, letter from the United Kingdom's Air Attaché in our Washington embassy included the following statement: "Secretary of the Air Force, the Honorable Donald B. Rice, was to say the least incensed by the renewed speculation, and the implied suggestion that he had lied to Congress by stating that Aurora did not exist. As you will have gathered, the whole affair is causing considerable irritation within HQ, and any helpful comments we can make to defuse the situation would be appreciated."

The United Kingdom moved quickly to prevent this becoming a full-blown diplomatic incident, but there was certainly an irony here, because at the same time as the UK authorities were asking their US counterparts about Aurora the US authorities had asked the United Kingdom whether the RAF had some secret prototype program of some sort. It was clear that the question was prompted by awareness of some of the UFO sightings detailed in this chapter, and again, this raises interesting questions, given the US government's public position that there is no current official interest in UFOs and that there have been no official investigations undertaken since Project Blue Book was terminated in 1969. Clearly, *someone* in the US government was interested.

I can certainly confirm that as recently as 2012 a US Air Attaché in another country attended a UFO-related briefing that Air Force officials in the country concerned had arranged. I can also confirm (not least because I was present) that a closed meeting took place in Washington, D.C., in June 2011 at which the UFO phenomenon was discussed with a group of well-connected personnel, including a former presidential chief of staff and a former CIA director.

The US denials over Aurora mirrored the denials that had been given in the legal action that followed the Cash-Landrum encounter. But are the denials true? Access to classified information is determined by your security clearance and your "need to know," and while the former is cut-and-dried, the latter can be the result of a much more subjective process. Furthermore, deliberately cutting people out of a project where they have both the requisite security clearance *and* the "need to know" is sometimes a deliberate tactic, designed, for example, to protect senior figures within an organization by giving them what's known as "plausible deniability." This makes it extremely tricky even for insiders to discover the truth. In his April 22, 1993, briefing to the Assistant Chief of the Air Staff about the

Cosford incident my head of division wrote: "If there has been some activity of US origins which is known to a limited circle in MoD and is not being acknowledged it is difficult to investigate further."

One can only speculate, but handing off certain black projects to a private corporation moves them outside the government (and thus outside the scope of the Freedom of Information Act) and means that the government denials would be true and given in good faith—provided that the person doing the denying wasn't in the loop!

In a final twist that would doubtless have infuriated the Americans even further, the Condign Report (the final report of which was dated February 2000) stated: "Some UAP reports can be attributed to covert aircraft programs." While Aurora was not specifically mentioned, it was certainly something that I know had been in the author's mind.

Not for the first time with significant UFO cases, there's a bizarre coincidence with the date. The UFO encounters collectively known as the Cosford incident took place three years (to the very night) after the UFO sightings in Belgium that led to F-16 jets being scrambled to intercept the unidentified aircraft that had strayed into Belgian airspace. Moreover, the description of the UFO—a large, triangular craft—was almost identical. It was for this reason that I asked our Air Attaché in Brussels to reach out to the Belgian Air Force to exchange information about our respective sightings. This was done, though again, this information was not made public at the time, as it would undermine the impression that we were trying to give, i.e., that we had little or no real interest in the subject.

The Cosford incident remains unexplained to this day.

NEAR MISS OVER THE PENNINES

On January 6, 1995, a Boeing 737 on approach to Manchester Airport nearly collided with a UFO over the Pennines at a height of four thousand feet. Both the pilot and the first officer saw the illuminated, wedge-shaped UFO pass down the right-hand side of the aircraft at high speed. Neither man was certain how close the object had come to colliding with the aircraft (which was carrying sixty passengers), but the first officer instinctively ducked as it flashed past. The MoD and the CAA launched investigations, but no explanation was ever found. The nervous post-incident exchange of conversation between the aircraft and Manchester Air Traffic Control authorities was recorded and reads as follows:

B737: C/s we just had something go down the RHS [right-hand side] just above us very fast.

Manchester: Well, there's nothing seen on radar. Was it, er, an ac [aircraft]?

B737: Well, it had lights; it went down the starboard side very quick.

Manchester: And above you?

B737: Er, just slightly above us, yeah.

Manchester: Keep an eye out for something, er, I can't see anything at all at the moment so, er, must have, er, been very fast or gone down very quickly after it passed you I think.

B737: OK. Well, there you go!

THE PHOENIX LIGHTS

In 1997 a wave of sightings took place over the city of Phoenix, Arizona, with hundreds of witnesses. Video footage was taken. The witnesses reported a series of lights in a v shape—either separate objects or maybe lights fixed to the underside of a single vast boomerang-shaped craft. To defuse the mounting panic and hysteria the Governor, Fife Symington III, held a press conference and had his chief of staff dress in an alien suit. Years later Symington—a former military pilot himself—confessed that he, too, had seen the UFO and apologized for the stunt.

UFO AT THE AIRPORT

On the afternoon of November 7, 2006, pilots and airport employees at O'Hare International Airport in Chicago saw a gray, disk-like object hovering over the tarmac for several minutes. The witnesses said that it hovered for several minutes before shooting off at high speed. A particularly striking feature of the sighting was that the object departed with such speed that it apparently punched a hole in the cloud cover, at a height of around two thousand feet. Because nothing was tracked on radar, the FAA did not investigate the sighting, despite the obvious air safety issue.

THE CHANNEL ISLANDS UFO

On April 23, 2007, Ray Bowyer, a commercial airline pilot, saw a massive yellow cigar-shaped UFO in the vicinity of the Channel Islands, between the United Kingdom and France. Some of his passengers saw it, too, as did at least one other pilot in the area. Air Traffic Control confirmed that the object was briefly tracked on radar but categorized it as "unknown traffic." Bowyer said that he wanted to chase the UFO but that the safety of his passengers had to come first. The MoD investigated the sighting but determined that the UFO had been in French airspace and was outside their jurisdiction.

UFO IN NEAR MISS WITH POLICE HELICOPTER

A spectacular encounter between a UFO and a police helicopter took place on June 8, 2008, in South Wales, over the military base at RAF St Athan, close to the Cardiff international airport. The helicopter, with a crew of three people onboard, was about to land when it was in a near collision with a UFO. Initially describing it as being disk shaped and covered in lights, early media reports suggested that a chase had taken place, with the helicopter pursuing the UFO south over the Bristol Channel and only breaking off pursuit when the UFO proved too quick and when they ran low on fuel. After the story broke on June 20, the account of events changed and the police were careful to use the phrase "unusual aircraft," rather than "UFO." In addition, while confirming the sighting, they denied that a chase had taken place. Perhaps the most extraordinary aspect of the story was a quote from the MoD Press Office, where a spokesperson made the following comment: "But it is certainly not advisable for police helicopters to go chasing what they think are UFOs." The advice was a virtual restating of Project Condign's "No attempt should be made to out-maneuver a UAP during interception" recommendation, though, as we have seen, the flight safety recommendations in the Condign Report do not appear to have been formally promulgated to the relevant aviation authorities, despite the DIS recommendation that this should be done. It's as if, out of embarrassment perhaps, an official somewhere said to himself, "We have a flight safety issue here, but as it relates to UFOs, I'll look foolish if I pass it on."

I have not attempted an analysis of the individual UFO sightings presented in this chapter. It is unlikely that a single neat solution could explain all the various incidents presented here. What I have attempted to do is place into the record a brief overview of some of the cases that most impressed me and other government officials charged with undertaking official research and investigation into the UFO phenomenon. In doing so, I think a number of points become clear. First, as with the Rendlesham Forest incident, conventional explanations are unlikely. Second, the US and the UK governments (among others) take this issue seriously and still investigate, despite assurances to the contrary. Third, in order to give plausible deniability to the claims about not investigating UFO sightings and to take the issue outside the scope of the FOIA and further away from congressional/parliamentary scrutiny, the issue may have been handed off to some unofficial cabal of officials within the defense/intelligence community or to a private corporation. As a former government official who has been involved in this activity I am reluctant to use phrases like "UFO cover-up," because they are constantly bandied about by the quirkier members of the UFO and conspiracy theory communities. But these communities have a point, as the evidence presented in this chapter shows.

There are lots of disparate elements to this and it's only when they are pieced together that a clearer picture emerges.

If all this seems a little heavy, then I shall conclude this chapter with one of the more amusing assessments to emerge from the DIS. It goes back to the reasons we set up Project Condign and a 1995 document that read, in part: "If the sightings are of devices not of the earth then their purpose needs to be established as a matter of priority. There has been no apparent hostile intent and other possibilities are: 1) Military reconnaissance; 2) Scientific; 3) Tourism."

Despite the lighthearted speculation, the intent was more practical, as the document went on to set out: "We could use this technology, if it exists."

16. OTHER VOICES

We have already heard how outspoken one former UK Chief of the Defence Staff (the UK post broadly equivalent to the Chairman of the Joint Chiefs of Staff) has been in relation to the Rendlesham Forest incident. Lord Peter Hill-Norton wrote to the MoD many times about the incident after his retirement and raised the issue several times in Parliament, as we have seen. Of all the letters he sent, few articulate his views better than the one he sent to Lord Gilbert, Minister of State, at the MoD on October 22, 1997. In addressing (or, rather, blowing out of the water) the MoD's untenable "no defence significance" assessment of the incident, Lord Hill-Norton summed up the issue as follows:

> *My position both privately and publicly expressed over the last dozen years or more, is that there are only two possibilities, either:*
>
> *a. An intrusion into our Air Space and a landing by unidentified craft took place at Rendlesham, as described.*
>
> *Or*
>
> *b. The Deputy Commander of an operational, nuclear armed, US Air Force Base in England, and a large number of his enlisted men, were either hallucinating or lying.*

Either of these simply must be "of interest to the Ministry of Defence,"
which has been repeatedly denied, in precisely those terms.

In this chapter, we shall look at some of the other opinions that have been offered on the Rendlesham Forest incident, by people whose opinions are relevant or particularly significant, given their position.

Before moving on, I should say a few words about Lord Hill-Norton. I first heard of Lord Hill-Norton's interest in UFOs through Timothy Good, a UFO author and researcher whom I first met in the nineties. Good is an unusually erudite and sophisticated ufologist who had, for many years, been a professional violinist with the Royal Philharmonic Orchestra and the London Symphony Orchestra. At first I was skeptical that Good is on friendly terms with a former Chief of the Defence Staff, but when I checked I found to my astonishment that not only was this true, but also Hill-Norton had penned the foreword to a book on UFOs that Good had written. Hill-Norton was clearly a believer in extraterrestrial visitation and, even more surprisingly, had come to believe that the US and the UK governments knew (at some level) about this and were actively covering it up. This put me in a difficult position in my dealings with Lord Hill-Norton. While I was the MoD's SME on UFOs, I was at the junior managerial grade, and Lord Hill-Norton was a five-star military officer! Technically speaking, he was not retired, because though he had long since stepped down from his post as Chief of the Defence Staff (and later as Chairman of NATO's Military Committee), five-star officers stay permanently on the "Active List" in the United Kingdom, so that in time of war they can be recalled to serve, should the current service chiefs be killed. Moreover, Lord Hill-Norton had a ferocious reputation for biting the heads off junior officers who irritated him. It was with some trepidation that I placed my first telephone call to him, in response to a request for a briefing on some matter or another, the specifics of which I do not recall. When he answered the telephone with a terse, "Yes?" I slipped into overly rank-conscious politeness: "Oh, good morning, Lord Hill-Norton. Nick Pope here. How are you this morning, sir?"

I was promptly cut off by his yelling down the phone: "Get on with it!"

After a quick, "Yes, sir, sorry, sir," I briefed him, and he must have forgiven my unsteady start, because we stayed in touch long after I left the MoD's UFO project. Indeed, he often asked me to draft letters for him on UFO-related matters—something that I undertook right up until his death in 2004.

It might be supposed that if there was some big secret about UFOs, known only to a select few, a former Chief of the Defence Staff might be one of those entrusted with the secret. Hill-Norton was adamant that the subject had never arisen while he was Chief of the Defence Staff, and I had no reason to doubt this. His frustration—and indeed anger—with the responses he got from the MoD when he raised the issue seemed genuine. It was with regret that he told me he'd never once thought to raise the issue proactively while he was in post.

Though Lord Hill-Norton seemed to favor the extraterrestrial hypothesis when it came to UFOs, he also lent his support to an initiative called UFO Concern, which was the brainchild of the Reverend Paul Inglesby. Inglesby favored a demonic explanation of UFOs, partly on the basis of the biblical description of Satan as being the "prince of the power of the air" (Ephesians 2:2).

Burroughs and Penniston are certainly grateful to Lord Hill-Norton. Penniston has this to say about the man and his efforts in relation to Rendlesham: "I am grateful that such a man existed at the time to help enquire into the Rendlesham Forest Incident. Unfortunately, he too was stonewalled, because of compartmentalizing. An honorable man of integrity and absolute resolve. I think Lord Hill-Norton was dismayed over the fences which lay before him. It was total frustration for him. However, I am grateful he championed the case at his level."

Burroughs wonders whether there was more to it than this: "It's possible Lord Hill-Norton had some insight into what was really going on inside the MoD with the phenomenon and wanted it to finally come out. He may not have known everything that went on, but he knew enough to know what was being said was not true and wanted something done about it; and Rendlesham clearly tied into the research the MoD did on Project Condign."

Whatever the truth, Burroughs was clear that such a thing could never happen in America: "You never see people that high up in the US military break ranks and talk about things that are classified because of national security."

Though Lord Hill-Norton's lobbying never led to any change in government policy on the issue, we have seen that his PQs and his letters certainly smoked out some interesting responses. In addition, for a field often portrayed by the media as being full of crackpots, it was enormously helpful for the UFO community to be able to show that one of its champions was a former Chief of the Defence Staff. For American readers, the situation would be analogous to somebody like Colin Powell speaking out positively on the issue and then accusing the DoD of covering up the truth! That might make a few headlines! The only surprise is how little media attention Lord Hill-Norton's outspoken views on UFOs gained in the United Kingdom. The explanation, ironically, is that belief in such things is so commonplace in the British establishment that it was hardly deemed noteworthy. Senior figures in the Royal Family and the senior ranks of the military and the intelligence community have long been interested in the occult and the paranormal, attending séances and studying subjects as varied as ghosts, UFOs, and crop circles. Figures who have taken an active interest in UFOs include Prince Philip, Earl Mountbatten of Burma, Lord Dowding, Sir Peter Horsley, and many others. The list of names is much longer, of course, but there are considerable sensitivities over this, for obvious reasons. Despite all this, it is only in the United Kingdom, perhaps, that a former prime minister (Arthur Balfour) could agree to become the President of the Society for Psychical Research. After all, the United Kingdom is a country where in 1979 the House of Lords (one of the two Houses of Parliament) held a three-hour debate on UFOs.

In 1997 the United Kingdom's former prime minister Margaret Thatcher (subsequently Baroness Thatcher) was caught up in the controversy over the Rendlesham Forest incident. The instigator of this was the irrepressible Georgina Bruni, the author and journalist who would later secure the release of the MoD's file on the Rendlesham Forest incident (though not, of course, the DIS file[s], which were mysteriously destroyed) by using the Code of Practice on Access to Government Information. Bruni was very well connected in political and diplomatic circles, and in 1997 she attended a charity dinner in London where Baroness Thatcher was the guest of honor. Toward the end of the evening, the two women fell into conversation, at first discussing the rising importance of the Internet. Bruni had started one of the United Kingdom's first online magazines and Baroness

Thatcher was always seen as a champion of the business community, though she was somewhat nervous about computers and the Internet herself. After a while, Bruni turned the subject of the conversation to UFOs and mentioned that she had spoken to a number of US military personnel who had seen UFOs or been involved in researching or investigating the phenomenon. Bruni specifically mentioned the Rendlesham Forest incident. Asked to comment on this, Thatcher retorted, "You can't tell the people." Bruni pressed the point and Thatcher rose to her feet and exclaimed, "UFOs!," causing her security detail to move forward. Bruni pressed the point. Thatcher offered the following statement: "You must get your facts right." Bruni tried a final time, but Thatcher amalgamated her two previous points and restated her view: "You must have the facts and you can't tell the people." At this point Bruni realized she would get no further, took Thatcher's hand, and thanked her for her time.

So what exactly did Baroness Thatcher mean by these comments? The question has been debated by ufologists and conspiracy theorists at either end of the belief spectrum. "You can't tell the people" is one of those cryptic remarks that has two meanings in this context. On one hand, as believers say, it might mean that Thatcher was privy to some great secret about UFOs (and maybe specifically about the Rendlesham Forest incident) that she strongly believed should not be made public. The reason for this might be because it was highly classified or perhaps because it would cause shock on a societal scale. The idea that there would be "panic in the streets" if the government was to disclose that extraterrestrials had visited Earth is one of a number of clichés that the UFO community is wedded to, though such a view is unsupported by any scientific research.

Jim Penniston has this to say about Thatcher's remarks: "A slip of the tongue; an oops moment; she knew some things and she caught herself when questioned with that response."

John Burroughs is more forthright on this point: "It's hard not to believe she did not have any insight into what the MoD was working on and why it needed to remain classified. She was saying you have to get your facts straight—which no one has—and if you do, you can't let the cat out of the bag because of national security. Again, it's clear, based on the declassified documents, that there's a race to weaponize the phenomenon—and you can't let any of that out."

Baroness Thatcher was Prime Minister at the time the Rendlesham Forest incident took place and was to become President Reagan's closest political ally. Could Reagan and Thatcher have been privy to a secret about UFOs and might the Rendlesham Forest incident have been part of this? Reagan had seen a UFO himself, on two separate occasions, while serving as Governor of California. Later, as President, he caused a sensation in a 1987 address to the United Nations, when he said, "I occasionally think how quickly our differences worldwide would vanish if we were facing an alien threat from outside this world." He made a similar point to the Soviet leader Mikhail Gorbachev, leading to a wide range of conspiracy theories suggesting that Reagan was trying to hint at the truth about UFOs. It has even been suggested by some that the real purpose of President Reagan's Strategic Defense Initiative (popularly known as "Star Wars") was to lay the groundwork for building a defensive capability against hostile extraterrestrials.

Skeptics interpret Thatcher's remarks in a more prosaic way, suggesting that she simply meant that if people want to believe in UFOs no rational, skeptical arguments will dissuade them from their dogmatic point of view, making sensible discussion and debate with such people impossible.

The second part of Baroness Thatcher's remarks to Bruni, "You must get your facts right" / "You must have the facts" is less controversial and according to those people who were close to Thatcher is the sort of thing she often said. As Prime Minister, particularly during the political cut and thrust of debate during Prime Minister's Question Time, Thatcher was well known for her mastery of complex briefs. She was a voracious reader and was well known for staying up late into the night reading every word of reports where most others would have read only the executive summary. Being able to recall accurately the details of an issue was one of the ways in which she dominated her colleagues and officials, so her comment about the importance of having accurate facts is arguably just a statement of her personal philosophy.

Georgina Bruni and Margaret Thatcher met on one further occasion, in the summer of 2006, at a social function on the terrace of the House of Commons. I do not know whether UFOs were discussed on this occasion, though a photograph taken of the two women at the time shows them smiling as they conversed, suggesting that Bruni's raising the topic of UFOs and the Rendlesham Forest incident with the former prime minister had not resulted in her falling from favor or being dropped from the guest lists of such high-powered functions.

Georgina Bruni died in 2008 and left no further records of her dealings with Baroness Thatcher on these matters. Baroness Thatcher never formally commented on her conversation with Bruni and died in April 2013. The precise meaning of her remarks on UFOs and on the Rendlesham Forest incident remains unclear.

Georgina Bruni raised the question of what happened in Rendlesham Forest with at least one other senior politician, Michael Portillo. Portillo had been Secretary of State for Defence between 1995 and 1997. When Bruni asked him about the incident in 2000 he confirmed that he was aware of it but pointed out that it had taken place before he had been appointed as Defence Secretary. Bruni pressed him, saying she was sure he had been briefed on the incident and asking if there was anything he would care to tell her either about Rendlesham or about the UFO phenomenon more generally. He smiled and said, "I know a lot, but I tell a little." Such quips can be endlessly debated by the UFO community, but while they might imply that Portillo possessed some information about the incident that he was not prepared to share, it could equally be the case that this was a lighthearted remark designed to fob off a persistent journalist and perhaps have a little fun with her at the time. Politicians—even at this senior level—are not averse to such things.

As the Base Commander at the time of the incident, Colonel Ted Conrad was second in command of the twin bases, outranked only by the Wing Commander, Gordon Williams. Conrad was Halt's boss and the person who, when the two of them were told that the UFO had returned on the evening of December 27, decided that Halt should be the one to go out to investigate.

Conrad has seldom commented on the Rendlesham Forest incident, but a UK newspaper, *The Telegraph*, quoted him on August 6, 2011. He was extremely skeptical: "We saw nothing that resembled Lt Col Halt's descriptions either in the sky or on the ground. We had people in position to validate Halt's narrative, but none of them could."

He went on to set out some thoughts on the incident itself:

The search for an explanation could go many places including the per-petration of a clever hoax. Natural phenomenon such as the very clear cold air having a theoretical ability to guide and reflect light across great distances or even the presence of an alien spacecraft.

If someone had the time, money and technical resources to deter-mine the exact cause of the reported Rendlesham Forest lights, I think it could be done. I also think the odds are way high against there being an ET spacecraft involved, and almost equally high against it being an intrusion of hostile earthly craft.

Conrad's most forthright comments were saved for Halt, with specific reference to his allegation of a cover-up: "He should be ashamed and em-barrassed by his allegation that his country and England both conspired to deceive their citizens over this issue. He knows better."

Halt was furious and fired a letter back to the newspaper. It stated in part:

Ted Conrad is either having memory problems, has his head in the sand or [is] continuing the cover up. . . .

Thru the years Conrad has made conflicting statements about the events . . .

Now he's smearing those involved. It's pretty clear there was a very intense confrontation with something in the forest. Does Conrad want to talk about how the airmen were then subjected to mind con-trol efforts using drugs and hypnosis by British and American au-thorities?

Clearly there is more going on here than a mere disagreement over the facts or the interpretation—there is obviously huge personal animosity between Conrad and Halt. Likely, this tension existed long before the Rendlesham Forest incident. Are we dealing with two very different indi-viduals who would have perceived and reacted to the events (and the af-termath) in totally different ways, with Halt in the role of believer and Conrad as the skeptic? Or is this unfair to Halt, who experienced all this in a way that Conrad did not? Had Conrad made the decision to go out into the forest and ordered Halt to remain at the awards ceremony would their

OTHER VOICES / 217

roles now be reversed? Does Conrad's scathing criticism of Halt reflect irritation at his perceived disloyalty, or is there a twinge of bitterness and
regret there? In my official government research and investigation into
UFOs, I have often heard witnesses tell me they would rather the event
had not taken place. Conversely, many people tell me that they would *love*
to see a UFO. Could the ultimate irony here be that Halt regrets being
caught up in the Rendlesham Forest incident, while Conrad wishes that he
had been?

While Penniston has no particular view on this aside from wondering
whether recollections differ over time and that "something personal" is
involved, Burroughs believes it goes further than this and recounts a meeting he had with Conrad, in recent years, to try to nail down what happened. Burroughs recalls that Conrad mentioned Occam's razor in relation
to the events, i.e., that the simplest explanation is usually the correct one.
But even when pressed twice, Burroughs recalls that Conrad never once said
Burroughs was wrong. As Burroughs relates, Conrad told him, "You have to
look at my standpoint; I was there to protect the senior officer [Williams].
That was my job—and still is."

As Burroughs summarizes the situation: "Colonel Conrad has admitted
an incident took place. What he's not willing to do is say what he feels it
was. He's not happy that Colonel Halt has been so open with what he
knows and in fact he told me personally Colonel Halt had broken the [in-
formal] officers' code of silence by making statements he's not authorized
to make."

Paul Hellyer is the former minister of defence and deputy prime minister of Canada. Amazingly for such a senior political figure (he is also a
member of the elite Privy Council), he not only is interested in UFOs but
also is a firm believer that some are extraterrestrial and that there is—at
some level—an official cover-up on the issue. I have met him on several
occasions, and he has also met Charles Halt, specifically to discuss the
Rendlesham Forest incident. When asked, Hellyer set out his views on the
matter as follows:

*The Rendlesham Forest case is an action thriller from start to finish.
I refer to it as the absolute classic of UFO sightings because it contains
all the elements and high drama that one only gets glimpses of in most*

stories. All that, plus authentication by several totally reliable witnesses. First, a sighting by young non-commissioned officers that was greeted with the usual incredulity and much laughter; later a second sighting reported by a "white faced" officer, results in the deputy base commander being designated to investigate and, in his words, "put an end to this nonsense once and for all." Instead, what he saw changed the life of this tough, experienced Air Force Colonel forever—a mesmerizing story that is stranger than fiction.

We live in a world where the word "hero" is overused and misused. But there are some genuine heroes out there and Dr. Edgar Mitchell is one of them. This former Navy captain and astronaut is also a member of one of the most exclusive "clubs" imaginable—he is one of only twelve people ever to have walked on the surface of the moon. Stories about astronauts encountering UFOs are for the most part spurious or based on misinterpretations of things like debris from booster rockets. Apollo 14 lunar module pilot Edgar Mitchell has not seen a UFO himself but is a firm believer that UFOs are extraterrestrial and that elements within the US government are aware of this and are involved in disinformation—if not an active cover-up—to hide the truth about this. He believes that this truth includes confirmation that an extraterrestrial spacecraft did indeed crash at Roswell in 1947. Mitchell's beliefs come in part from things that he has been told by sources within government, the military, and the intelligence community. As he was an Apollo astronaut and all-American hero, clearly these are communities where Mitchell moves freely, at the highest levels. When I asked Dr. Mitchell about the Rendlesham Forest incident, he told me this: "The Rendlesham Forest UFO incident is important in that it occurred near a US military encampment in the UK, one not noted for interest in the UFO phenomenon. It was observed and reported by a number of personnel independently and consistently reported."

Most people have an opinion on UFOs, and those opinions tend to be polarized, with some believing just about everything and others being deeply skeptical. Previously, Roswell has been the battleground on which this debate has played out, with believers championing the theory of a crash of an alien spacecraft, while skeptics push the theory that it was just a weather balloon. Now that Roswell has passed from living memory into history, Rendlesham Forest is the new battleground, with a more diverse bunch of theories for believers and skeptics to grapple with. But for anyone wanting to portray UFO believers as crackpots and kooks, Rendlesham poses a problem, because its proponents include not just the military personnel involved (compelling though their testimony is) but also senior political and military figures whom one would not normally associate with this subject.

But endorsements and opinions are not enough. The witnesses to the Rendlesham Forest incident may be glad that they are believed at a high level, but all this counts for nothing in the absence of the official confirmation that they seek.

17. THE SEARCH FOR ANSWERS

When we began to research and write this book, we realized that it would be highly advantageous to use the Freedom of Information Act to seek information about the Rendlesham Forest incident from various US government agencies. It was fortunate, therefore, that John Burroughs and Jim Penniston were already in touch with a Mississippi attorney, Pat Frascogna, who had considerable experience exploiting the US Freedom of Information Act and taking on large bureaucracies.

If, as the evidence strongly suggests, the object that landed in Rendlesham Forest generated radioactivity that MoD scientists assessed as being "significantly higher than the average background," one can well understand why Burroughs and Penniston are even more eager than the other witnesses for an official explanation. Fortuitously, Pat Frascogna was an expert in the very areas that we needed to take forward the investigation. Even more fortuitously, he had already embarked on a quest to probe various agencies that might have information and/or records pertaining to the incident and to Burroughs and Penniston themselves. Moreover, because Frascogna had a personal interest in UFOs, he was (with the full consent of Burroughs and Penniston) prepared to share the results of his work and was prepared for the details to be published.

On most occasions, when writing a book, an author paraphrases and/or summarizes a vast amount of primary source material, quoting from it di-

rectly where most appropriate. However, with material written by an attorney, there are clear dangers for an author in attempting to summarize and/or paraphrase the text. In legal language, altering even a single word can have a dramatic effect on the meaning. Accordingly, what follows in this chapter—with the exception of some final concluding remarks—is written entirely by Pat Frascogna, so as not to alter the meaning in any way.

In 2008 I happened to see an episode of the TV show *UFO Hunters* titled "Military vs. UFOs." This episode was my first ever introduction to the Rendlesham Forest incident. Although I certainly had no idea at the time that episode aired, it was the beginning of a journey that would years later lead me to John Burroughs and Jim Penniston. The Rendlesham Forest encounter immediately struck me as very different from the vast majority of UFO sightings. I was so fascinated by the incident and its witnesses that I steeped myself in Internet sources regarding it. Three years subsequent I was ready to confront the principals of the story, John and Jim, and invited them to Mississippi to speak at the first UFO conference ever held in the state. In reality, I organized that conference just so I could meet these gentlemen, Linda Moulton Howe, and Tom Carey, a leading expert on the Roswell crash. It was from the 2011 conference I organized in Mississippi featuring John and Jim that they invited my participation with them to search for answers. As an attorney of twenty-two years and longtime local impresario, I was both honored and eager to volunteer my time and lend expertise to their cause. The decision to help these gentlemen was an easy one.

In 2011 and 2012 many FOIA requests were propounded on government agencies. These agencies included the Central Intelligence Agency (CIA), NSA, DIA, Department of the Air Force (DAF), DARPA, via the DoD, Department of State, AFOSI, and Department of Veterans Affairs. Considering the fact that John and Jim each served in the Air Force for many years, that they were stationed at RAF Bentwaters in 1980 during the height of the Cold War, and that they were entrusted with protecting military assets and advanced destructive ordnance would, by itself, be enough to assume that at least one of the agencies listed previously would "remember" them in their files. Add to those facts the events in

Rendlesham Forest in December of 1980 and it seems rather certain that John and Jim would have quite a paper trail of records of the time they served their country in its armed forces. Not the case if, that is, one is to believe the litany of phrases contained in the responses to our FOIA requests. The first round of FOIA requests in late 2011 made identical requests for both John and Jim:

> *I have been retained by [Penniston/Burroughs], formerly of the United States Air Force, who was stationed at RAF Bentwaters and RAF Woodbridge near Suffolk, England in December of 1980. An original Declaration signed by Mr. Penniston pursuant to 28 U.S.C. § 1746 is included herewith. Under the Freedom of Information Act, 5 U.S.C. Subsection 552, I am requesting any and all information and/or records, regardless of format, which name specifically and/or reference in any manner, my client in connection with RAF Bentwaters and/or RAF Woodbridge, and "lights," "plasma fields," "unidentified aerial phenomenon (UAP)," "craft," and/or "vehicle(s)" in the Rendlesham Forest, commencing December 25, 1980 through December 31, 1981.*

In early 2012, another round of FOIA requests was dispatched, again identical for each gentleman, and sent to all the same agencies as before:

> *I have been retained by . . . [Penniston/Burroughs], formerly of the United States Air Force (USAF), who was stationed at RAF Bentwaters and RAF Woodbridge near Suffolk, England in December of 1980. An original Declaration signed by Mr. . . . [Penniston/Burroughs] pursuant to 28 U.S.C. § 1746 is included herewith. Under the Freedom of Information Act, 5 U.S.C. Subsection 552, I am requesting any and all information and/or records, regardless of format, which name specifically and/or reference in any manner, Mr. . . . [Penniston/Burroughs] in connection with his service in the USAF while stationed at RAF Bentwaters and/or RAF Woodbridge, specifically, during the time period commencing December 25, 1980 through December 31, 1981.*

The responses we received to both rounds of FOIA requests were interesting but certainly not revealing of any specific information:

CIA: The CIA can neither confirm nor deny the existence or nonexistence of records responsive to your request.

NSA: A thorough search of our files was conducted, but no records responsive to your request were located.

DIA: The Defense Intelligence Agency has determined that any responsive files would have been transferred to the National Personnel Records Center.

DOD: A search of the records systems maintained by the Defense Advanced Research Projects Agency revealed no records responsive to your request.

DAF: After a thorough review of your request, it is determined that the requested records are no longer under our purview.

AFOSI: We have conducted an extensive search of our files and ran Defense Central Index of Investigations (DCII) database checks and other various AFOSI databases checks. It has been determined that the Air Force Office of Special Investigations is not maintaining any information responsive to your request.

State Department: Some or all of the records you have requested do not appear to be State Department records (other agency [Department of the Air Force] information may be attached).

In other words, none of the agencies quoted seem to know anything about a John Burroughs or a Jim Penniston, despite the fact they were most certainly in the USAF many years.

In addition to the FOIA requests we propounded, efforts have been made to obtain John's and Jim's medical records from their years in the service, specifically coincident with their time at RAF Bentwaters in 1980. These records would, perhaps, reveal what it was John and Jim were exposed to in Rendlesham Forest in 1980. Physicians of John's, especially, have requested he obtain such records to aid in their assessment of the health problems he has suffered ever since his encounter in the forest. This has become a very serious request, indeed, that professional medical personnel have repeatedly asked John to provide them.

Unfortunately, this has proven to be a particularly aggravating quest. The reader should recall that these medical records were requested as part of every FOIA request issued for which none, even from the DAF, were produced in response thereto. However, John's and Jim's medical records have been sought separately, as well. Letters making requests and agency forms filled out by John and Jim have netted little more than statements like "When we have completed our search for records responsive to your request, . . . [we] will send you another letter telling you the results of that search and our next step in processing your request," only to never hear anything from the Department of Veterans Affairs again. John Burroughs has even tried obtaining his medical records with the assistance of former Arizona senator Jon Kyl, who, after almost a year, had no luck whatsoever in getting John's records. Senator Kyl did, however, through the many months he tried to help John, reveal to him that there is a classified-records section of Veterans Affairs (VA). Evidently, John's records are classified, but why they would be is unknown.

The first ten years of my legal career were spent chiefly in criminal defense work, including three years as a full-time public defender. The decade-plus since then I have primarily been engaged in consumer protection lawsuits against well-funded corporate defendants. I have been exposed to and am very familiar with institutionalized deceit of corporate defendants in their business practices. Moreover, the compartmentalization of the deceit in corporate defendants I have seen produces many "left hands" never knowing that the "right hands" even exist, let alone what role in perpetuating the deceit they play. The responses we received to John's and Jim's FOIA requests remind me exactly of this. As is obvious by the previously quoted responses to our FOIA requests, nobody claims to know or have anything to produce, not just regarding the Rendlesham Forest incident but regarding John or Jim, either! For example, the Air Force was asked if they had "[any] information and/or records, regardless of format, which name specifically and/or reference in any manner, Mr. Penniston in connection with his service in the USAF while stationed at RAF Bentwaters and/or RAF Woodbridge." The response we received to that seemingly simple question was that any such records "are no longer under our purview." Contrast that ridiculous response with that of the State Department when asked the same question; they responded by advising us to ask the Department of the Air Force and provided us with contact information for same. As if those responses were not spurious enough, then it was the response we received from the CIA that is the most tantalizing. Their use of the Glomar response or denial, "can neither confirm nor deny," raises the question that if they have nothing in their files on the Rendlesham incident, then why not just say so? In addition to invoking the Glomar denial, the CIA has further stated in their response:

> The fact of the existence or nonexistence of requested records is currently and properly classified and is intelligence sources and methods information that is protected by section 6 of the CIA Act of 1949, as amended, and section 102A(i)(1) of the National Security Act of 1947, as amended. Therefore, you may consider this portion of the response a denial of your request pursuant to FOIA exemptions (b)(1) and (b)(3), and [Privacy Act] exemptions (j)(1) and (k)(1).

Section (b)(1), as referred to here, reads "exempts from disclosure information currently and properly classified, pursuant to an Executive Order."

Section (b)(3), also referred to here, reads "exempts from disclosure information that another federal statute protects, provided that the other federal statute either requires that the matter be withheld, or establishes particular criteria for withholding or refers to particular types of matters to be withheld. The (b)(3) statutes upon which the CIA relies include, but are not limited to, the CIA Act of 1949."

Sections (j)(1) and (k)(1) of the Privacy Act, as cited by the CIA, read, respectively, as follows: "exempts from disclosure certain information maintained by the Central Intelligence Agency" and "exempts from disclosure information properly classified, pursuant to an Executive Order."

Thus, according to the CIA, our FOIA request for "any and all information and/or records, regardless of format, which name specifically and/or reference in any manner Mr. Penniston [or John Burroughs] in connection with his service in the USAF" is denied because disclosure of that information would violate an Executive Order and federal statute protects such information from disclosure.

The governmental response to our FOIA requests is essentially no different from what I have seen countless times in suing corporate entities. First, there are many strata that comprise the corporation/agency/et cetera. These layers typically operate independently of the other, nor do they really have access to what the others do. Some might properly call this compartmentalization. Second, despite the multiplicity of layers, there is a grand design to them. They are each part of an overall scheme or purpose, and they always have a built-in defense mechanism to stop intruders. Thus, whether it be learning the dirty and unethical business practices of a company or the secrets of our government, the same deployment of denials and feigning ignorance about what is really going on are the all-too-common methods used to keep the truth from the light of day.

Pat Frascogna's report, which was reproduced verbatim, here, is only an executive summary of a lengthy and complex effort on his part. In view of the unsatisfactory nature of the responses received to date, work is ongoing and further attempts are being made to secure acceptable answers from the various agencies concerned.

We do not intend to comment at length on the material presented in this chapter. We should, however, make one point. In cases where files have been shredded or documents lost, there is always a debate to be had over whether such things are indicative of a conspiracy or whether it might simply be a case of bureaucracy. It is certainly the case that in any large organization files and documents go missing or staffs, for whatever reasons, are unable to locate them. However, in government, the military, and the intelligence agencies document security and information management are generally taken extremely seriously, for obvious reasons, especially where classified and/or sensitive operations are concerned.

How can we decide whether this is conspiracy or bureaucracy? First, readers will recall that this is not the first time information relating to the Rendlesham Forest incident has gone astray. On some of the nights at the time of the incident, the radar cameras were apparently turned off; when they were turned on, the films were fogged; several people who took photos of the

UFO (including Jim Penniston) were subsequently told that the images had not come out; most critically, DIS UFO files covering the period of the incident were destroyed, without anybody being able to say why or by whom. At the very least, one can say that a pattern is beginning to emerge and the difficulty Frascogna has had securing documents and medical records certainly fits this pattern. Second, Frascogna is an experienced and tenacious lawyer, well used to taking on large and powerful organizations. If the sorts of responses he received suggested simply an administrative error, he would say so. He has not. His experienced assessment is that something else is going on here.

Where does this leave us? Frascogna's efforts are continuing and help will be enlisted from other political figures in an attempt to secure access to the relevant records. In the United States, the military are highly regarded, whether they are serving members or veterans. Long gone are the dark days when returning veterans of the Vietnam War were treated so shamefully. Veterans of more recent wars, in Iraq and Afghanistan, are rightly regarded as heroes. The oft-used phrase "Thank you for your service" is said with genuine feeling. Against this background, it is puzzling and troubling that Frascogna has not been able to secure the medical records of two ex-service personnel who believe they are suffering adverse health as a result of an incident that took place while they were in the military.

John Burroughs and Jim Penniston will continue to work with Pat Frascogna, but the quest for a response from officialdom is only one half of the current story. Recently Burroughs and Penniston have become not just the focus of the ongoing campaign for answers about what the various military witnesses saw and experienced but also the two people at the center of a storm threatening to engulf the entire UFO, conspiracy theory, and alternative belief community. Because Burroughs and Penniston have recently revealed new details of what happened when they encountered the UFO on the first night. And these new details change everything.

18. THE RENDLESHAM CODE

Over the years, as bits and pieces of the Rendlesham story came to light, Jim Penniston kept one staggering aspect of his encounter secret. In chapter 1, we heard how out of all the witnesses who saw the UFO over three nights he came the closest. Indeed, he touched it. The secret he kept to himself, for over thirty years, is that *when* he touched it something extraordinary happened. Essentially, Penniston claims that when he touched a particular symbol he received a sort of "telepathic download" of ones and zeros, which he now believes was a binary code message.

Here, in Penniston's own words, is what happened when he approached the craft:

> As I move along the left side of the craft, I see what appears to be symbols of some kind, which were coming into view, as I move closer, my heart is beating as if it would jump out of my chest, as I write down what I am observing. I am shaking, it is beyond control, and I cannot stop this shaking. I stretch my hand out and close it into a fist, trying to regain steadiness. It seemed to help a little. As I look down the side of the craft, I see the inscription, is like nothing I have ever seen before, no aircraft marking, or no writing that I can identify. I am in an immense feeling of being overwhelmed, stacked with a feeling of total disbelief. After quickly drawing the glyphs down in my notebook, I was ready for a second go-around of this craft-of-

unknown-origin, so I put my notebook back in pocket and initiated the second walk around on the 360, more concentrating on the detail of the craft, and the fact it seems to be defying gravity by the absence of observation of landing gear, or at least landing gear as I have known it to be. As I walked down the right side of the craft, I realized that I was still intact, and not harmed. I was making small moves of assessment, noting that no sign of threat or aggression was apparent to this point. I figured if I was to die, it would have happened by this point. A small feeling of relief enveloped me. So I decided to take it a step further, I decided to touch the outer skin of the craft. Or maybe I was compelled to do that, it was never clear to me. As I ever so slowly walked or shuffled, I took my hand and touched the craft momentarily, then I started to shuffle along while taking my hands and running them along the side of the craft. One hand over the other, as I moved around it. After almost a complete circle I come back to the symbols or glyphs if you will. Running my left hand from the smooth fabric to a course sandpaper feel of the glyphs, slowly feeling and tracing with my hand these pictorials. I run the bottom glyphs which measured about three feet long and a hand print high in height. Was the first one touched, then then I traced my finger tips on and then on to the next . . . then on to the last one in the line of glyphs, I moved my fingers over this one. Then above the hand wide height of these glyphs that ran along for about two feet or so, was a much larger one, one that was more intriguing, and one which seemed to take on more importance than the others did. It measured I will estimate a couple of hands high and about three hands wide. It was a large etched or embedded circle, with a large triangle running from top to bottom. With two smaller circles, one at the top of the triangle and one slightly larger one on the bottom right of the triangle.

I left not touching the large one, till last, it seemed like this was by design. I was curious but I felt that this was the key to the craft. I took my fingers and ran them slowly around the outer circle of the craft. I then took my fingers off, and hesitated while making a closed hand, and then I opened my hand and flatly pressed the palm of my hand on the large triangle. What was to happen next, defies reality. For I was no longer to see, what I was seeing in what I will call the mind's eye.

Was bright and steady brilliant light, I was squinting? Then the stream of ones and zeros ran relentlessly and I was unable to see my surroundings, I was scared, though I seemed to understand it was not harmful, but required. I am not sure how long this took place, this bright light with flashing of these ones and zeros. Seconds, or minutes, it seemed like a brief moment. Unable to pull my hand back, I finally had it release my hand. As quickly as the release of my hand from the triangle, the ones and zeros stopped. Thank God, I thought, I quickly started looking at my hand, I was examining it for damage. Seeing none, I was not going to do that again. No need to have physical contact again with the craft. Scary and unreal it all was.

I started to see the craft generating color through the fabric of the craft again. It started to become bright again, and began moving through the trees and then hovered at the top of the trees and with a flash and a blink of eye it was gone.

I hear movement to my right; it's John. I am wondering where he had come from and where was he during this encounter. My thoughts are broken with John quickly pointing to a direction out in the farmer's field. "There it is, Jim, let's go!" As quickly as he said that, he was off and I was right behind him.

So what happened with those flashes? Well, years later, I find that they are not just ones and zeros; I always had a thought or feeling that they meant something, but never acted on it. It is like all was supposed to wait. Like everything had its own time line. Or it could be something simpler, like a part of the condition of Post-Traumatic Stress Disorder (PTSD). Time kept travelling on, and then more than a decade passes and finally some answers about those ones and zeros. Apparently the binary codes were a direct result of contact with a physical craft, a craft of unknown origin. Meaning it was an unidentified craft and where it came from is still unknown.

The next day, while looking at my notebook . . . the glyphs in particular, I have had the codes running through my head since the incident the day before. I had a feeling to write them down . . . for I did and immediately after finishing them the codes were gone from my mind. I was finally at rest with them. The notebook was then put away and retired to a box. For a new one for work I had available. My thoughts

at the time, although profound, were actually much simpler. I wanted
it to go away, and I had no need to talk about it either. For this was
not to be the case.

It would be tempting to dismiss Penniston's claims out of hand. Some people will doubtless do so, and that is their right. However, it is not as simple as that, because Penniston's claims cannot be taken in isolation but must be considered as part of the wider series of events. The UFO he saw was seen by others, including the Deputy Base Commander. It was tracked on radar and left physical traces such as damage to the trees and—most tellingly—radiation levels that were assessed by the MoD as being significantly higher than background levels. Moreover, Penniston is hardly your average witness. At the time of the encounter he was a staff sergeant who was the on-duty flight chief at one of the most sensitive military installations in NATO. Because of the nature of his responsibilities, he had been formally evaluated and judged to be fully compliant with the USAF's PRP.

In writing this book, I had numerous questions for Penniston about this aspect of the story. Rather than write up an account based on our various discussions, I have made the decision to reproduce a lengthy statement that Penniston drew up for me in March 2013. I will make only a few minor, explanatory comments. This is a strategy familiar to me from my government work, where in the world of intelligence analysis the raw data is all-important. An example of this is the Iraq War, where the media and the public were told that Iraq possessed weapons of mass destruction—a questionable (and maybe political) analysis that was extremely difficult to reconcile with the raw data.

I think it is abundantly clear from what follows that Penniston is still traumatized and confused by these events to this day. His direct and frank writing style marks him out as somebody who shoots from the hip and I have not sought to tone this down:

The binary code or should I call it the binary enigma? To tell you
where it is now at, I think we must go back and look at how the ex-
change of ones and zero happen in the first place.
On night one or the 26th of December 1980 is where it had all
started and I could not have foreseen what the impact for me on this

would be. The binary codes were a direct result of my unexpected and humbling contact with a physical craft, during the first night. As I investigated the remarkable event, I was seeing things that were unexplainable at the time and for that matter are unexplainable today. The longer around the phenomena, the more confusing it all appeared. As each moment passed with a flow of unknowns that were unfolding before me. I was trying my best to evaluate what it was, and more importantly what it wasn't. I had asked myself: how could this be happening? I was trying to conceive a rational explanation, of what this could be. And I kept asking myself why was it sitting in front of me in this small clearing at a forest? As I observed, I knew that this technology was extremely advanced, far above any that I had seen to that time, and even to this day.

Thus leaving the only way to report is to keep it strictly in military terms. My military assessment (one that would be acceptable in my reports), a craft of unknown origin, simply stating it was an unidentified craft and where it came from is still unknown today with a 100 percent certainty. The communication of binary codes (the term "binary code" was unknown to me at the time and I did not make the connection until October 2010) was accomplished when I physically touched the craft's glyphs, which were located on the outside skin of the craft. It activated a technology which is unknown to me, and apparently to everyone else too. The technology then communicated a series of ones and zeros to me. The communication transfer was accomplished within minutes. There was an area of about fifteen feet which surrounded the outside of the craft. This area I will call the bubble. For within the bubble, static electricity pulsed upon my clothes, skin, and hair. Also an appearance of slowing of time. The air seemed dead, not transmitting any sound. An extraordinary and sensational event, for sure, one that defied all I ever knew, or all I could ever imagine.

The following day in the morning, at my then home in Ipswich, I finally had gotten up. After a very disrupted night of nightmares and very little sleep. During the night I was often wakening with thoughts, or rather images, of visions of ones and zeros running through my mind, my mind's eye. It was an overwhelming and scary experience for sure. I had thought it was something on the verge of madness. I was struggling and I tried to compose myself the next morning. The trauma of the night/day before was finally starting to begin setting in. As I sipped on my cup of coffee, I felt so alone. As I was at the dining room table, I was so perplexed by the happenings of the day before. Trying to make sense, but more so, hoping and wishing these numbers would leave my mind. Overwhelming at times, and annoying at best, the bigger worry was the fear I was somehow damaged or hurt from my contact with this thing. How can these ones and zeros seem to be imprinted in my mind like a hot branding iron? My only hope was they would dissipate as the hours and days would pass. Because there were no other options, for I was not going to the base hospital with this. If the incident was not a career ender, then reporting for medical treatment with a recount of ones and zeros keeping me awake and overwhelming my thoughts was surely not the way to go, and this I truly felt would be the end of my air force career.

As I sat there, and picked up things that would later become evidence of that night, I relooked at the pod casts that I had made shortly after the encounter by us some 24 hours earlier. As I sat there in somewhat disbelief I slowly ran my fingers over the landing pod cast, still somewhat damp from being poured shortly after the incident and the day before. Picking it up at times and smelling the mustiness of the plaster of Paris, its odor was with the distinct smell of the forest floor clearly on it.

As I reflected, my mind wondered in thought of what it could have been that me, Burroughs, and Cabansag saw and pursued at different points on that Friday night. If that was not enough, I was also thinking of the flashes and the flashing of the blinding light, and the craziness of the ones and zeros which seemed to flash across my sight, but only blind to all other things. And for God's sake, why could I still see them? Laying down the pod, I then picked up my notebook that I carried that night from the side counter where I had emptied my pockets from the day before after coming home from working my last midshift. I sat back down and began to thumb through the pages; I patiently relooked at the glyphs that I had copied down. While I had begun to start looking through the notebook for the third and then for the fourth time.

The glyphs in particular had my attention. As I did this, the ones and zeros started coming back to me, almost bothersome as they start up again; there I sat, having what had kept me up the night before to start again. I had the ones and zeros begin to start running through my head again. Not blinding as before, but flashing in what seemed a sequential fashion. I was so worried that this trauma and feeling I could never share, this madness, with no one. I was actually wondering if my sanity could survive this barrage. I was forever a prisoner of my own madness. Since the incident the day before, this was the new normal for me, so my thoughts were clearly were how was I going to survive this trauma that I have experienced? How could I even function with day-to-day life, not to mention curing myself from it? It was then that it happened, a momentary lapse of reason. A thought of writing those "ones and zeros" down. The more I thought about it, the more I felt compelled and was at the fringes of obsession to copy what my mind's eye was seeing. I had an even more overwhelming urge or, more precisely, I had a feeling of compulsion to write them down, this was now imperative and it had to be done immediately without further delay.

Finally I started by copying on the only available paper that I immediately had in front of me. It was my notebook. So I randomly flipped back to some new pages and so it began. So hesitantly I slowly started to write down the numbers, for I thought that this was beginning in the memory of madness. I first wrote a zero and then a one, and another, as I began to write it all down one number after another, as I was seeing the flashes in his mind. I wrote them down on a steady pace. As it progressed. Slowly as this act of madness was continuing. I was feeling a little better, a little bit that is. The small pieces of peace and fragments of satisfaction began. I started to write them down faster. I only had to stop a couple of times, because the pens I was using to record these ones and zeros kept stopping and apparently kept running out of ink. Each time that would happen I would momentary panic while seeking another pen from the drawer of the cabinet, look-

ing for another pen. Finding one, I would then sit back down at the table, and began my quest, my mission again to write those ones and zeros down, through this episode of madness of writing them down; I was finally seeing greater relief. I wrote effortlessly, as fast as they would flash in my mind, I would transcribe them to paper.

I did this continuously for about three-quarters of an hour or so. What began out of an act of desperation and madness ended as fast as it had begun. Upon finishing the writing of these numbers. What started so quickly just stopped. Seeing no longer any images of those ones and zeros in my mind. Immediately after not seeing them no more and stopping of writing them, it was then the flashing of the ones and zeros was gone in their entirety from my mind. For the first time since it began, some 24 or so hours earlier, I was finally at rest with them. An immense feeling of relief came over me. But something even worse stayed with my immediate thoughts. The thought of trauma and writing of these numbers like a madman. I asked myself: how can I tell anyone about this? The obsession to write down flashing numbers in my mind that were relentless since the encounter with glyphs on the craft. Who could I talk to? I surely could not go to the base clinic and complain about this? I would be looked at like a madman I was solidly convinced. After all I was certified under the Personnel Reliability Program (PRP), I was on the firearms issue and carry list. And of course I had my security clearance. I thought that everyone that I would tell about the incident, they would be obligated to report what I would say or had experienced. All the trappings of insanity were dripping all over it. Yep, most certainly it was a career-ender.

One of the questions I wanted answered was whether the issue of the binary codes was raised in any of the debriefings and interrogations that followed the incident. Penniston's statement addresses this:

The choice was clear, since writing those ones and zeros down, I was relieved that my mind was now clear and free of those things. Thoughts of what had tormented me from shortly after the incident and seemed to run unabated were actually gone. I was free, and more importantly I was relieved. The situation before was not a situation to deal with by myself. As the minutes went by, the clearer things had become. For what had happened here, it looked like I was going to be alright. I felt normalness for the first time since the incident. Never to mention the madness of the ones and zeros again to anyone, for they are now history and I am on my way to some normalness in my career. The ones and zeros are gone forever, or so I had thought, the madness of them would never be known. On the evening of the 28th of December 1980 just after supper I received a call from a Sergeant who worked in the Security Police orderly room. He told me that he received a call from the Squadron Commander to have me report to the AFOSI building in the morning. I told him I would comply. Funny thing is, he never asked any questions either.

Over the next few weeks and months I was methodically and consistently interviewed and interrogated by people within my chain of command, other Air Force agencies and others who were of other branches of service and other departments of the United States. Sometimes remembering those discussions and interviews and sometime not remembering them for decades. Every time, I was promised that this was the last interview and it would be absorbed into the classified annals of data and I would need to tell or talk about it no more. This was not the case. I went through at least fourteen debriefings and two by non–air force personnel. And many that I would not remember until decades later. I gave all information from memory and at no time was the notebook ever brought up. The debriefs were always the last one and were all for the last time; for if it was not AFOSI, it was to be everyone in my chain of command and all that I would meet with an inquisitive question. At the AFOSI building I was told and was promised: tell all and tell it correctly with no omission and it would be the last of questions on Rendlesham. How untrue this would become, as the years passed. For these were to continue, no matter what I had said. For it was unknown to me at that time, or the other visits to AFOSI or wherever it would be, I would not have memory of them, well, at least until them being told in the hypnosis some fourteen years later.

So, bizarrely, Penniston chose not to raise the issue. Though he does not say so, I think this goes further than the simple fact that he wasn't asked about it. This seems disingenuous to say the least, because when one is asked to give a full account of an event the omission of a germane fact is nearly as bad as a lie. It seems to me that either the memory had been suppressed or he chose not to raise it, for fear of losing his PRP certification or even being discharged.

The first time Penniston mentioned the codes was in 1994, shortly after he left the USAF. Suffering from health problems and sleep disorders, he had no belief that the sleep disorders were a result of anything other than stress. Penniston privately contacted a local hypnotherapist who could address the sleep disorders and the associated waking up throughout the night. It is important to say that she had nothing to do with the UFO community, nor did she even believe in the paranormal. After a few appointments it was agreed upon to do a regression hypnosis session and during the taped interviews (taped for the hypnotherapist's notes) it was discovered that a portion of time in Penniston's memory was blocked. She said, "It was the first time I have ever encountered this, although I have read about it." What came out of the regression astonished the hypnotherapist as well as Penniston.

Ufologists—particularly researchers into the so-called alien abduction mystery—often use regression hypnosis to try to recover hidden memories of extraterrestrial encounters. Ufologists believe these memories have been either suppressed by extraterrestrials or repressed by the individual concerned, as a self-protection mechanism, with the traumatic memories taken out of conscious awareness. We have previously highlighted the debate over the validity of regression hypnosis in recovering lost memories, but I present this information for readers to make up their own minds. The relevant part of Penniston's statement addresses some health issues before going on to cover the hypnosis session in which the binary codes are mentioned:

> In 1991 medical issues with my inner ear are now diagnosed as Menière Disease. In 1993 I retire from the USAF. In 1994 I have severe sleep issues and post-traumatic stress from the incident, and continuing problems from the now diagnosis of Menière Disease. The

Veterans Administration treated Menière Disease. They would not afford post-traumatic stress treatment for the incident. They said there was no evidence of anything happening at Bentwaters in 1980 which could produce this condition. So I had to seek private help for the sleep disorder and the post-traumatic stress. I also sought to remember, for I had memory voids of things from that time.

Transcript extract of my second hypnosis session on September 10, 1994, I begin to break through an apparent post-hypnotic suggestion by the AFOSI hypnotizers to keep me from talking. The following are excerpts from that second session in which the American and British intelligence people seem to have anticipated before their sodium pentothal hypnosis that I received a telepathic download of binary numbers.

EXTRACT OF HYPNOSIS SESSION, SEPTEMBER 10, 1994

Hypnotherapist: *You weren't supposed to understand the program?*

Penniston: *No. By touching these things [the raised glyphs on the craft's surface] I activated these things.*

Hypnotherapist: *You touched the symbols and you set off a program?*

Penniston: *Yes. It was repairing itself. All they wanted was a place to stay while it repaired itself.*

Hypnotherapist: *And by touching the symbols, you disrupted the repair program?*

Penniston: *I activated a binary code. The two [government agent] men want to know why.*

Hypnotherapist: *And what do you answer them?*

Penniston: *They ask me if I ever had any other encounters with them [binary and them the time travelers]. I haven't. They are discussing it between themselves. The situation. They've got a problem.*

Hypnotherapist: *What's their problem?*

Penniston: *Their problem is because I can't tell anybody. They ask no more questions about the craft. And they want to know [discussing with each other] what to do with me.*

This is a fascinating development. While theories about time travel have previously cropped up in ufology, they are uncommon and not popular. So if the hypnotherapist was somehow leading the witness or if Penniston was confabulating (giving false information honestly, e.g., as a result of brain damage or a psychological disorder), one would expect him to come up with a story about extraterrestrials, not time travelers. But yet he was very clear on this, at one point stating, "They are time travelers—they are us."

Penniston clearly had his own doubts about the time travel theory and about the binary codes. Indeed, he is at pains to point out that he did not originally connect the ones and zeros with binary code.

Once the possibility had been raised, Penniston shared a small part of the codes with some trusted individuals. These included Linda Moulton Howe, a filmmaker and journalist with a strong interest in UFOs, Kim Sheerin, co–executive producer on the popular History Channel TV show *Ancient Aliens*; Nick Ciske, a Web designer who is an expert in ASCII (a character encoder) and binary code; Joe Luciano, a computer systems engineer who had experience of binary codes while working in the USAF; and Gary Osborn, an author with an interest in a wide range of esoteric subjects.

Kim Sheerin recalls it like this: "In October of 2010, our production team spent a day with John, Jim, and Linda in Phoenix interviewing each of them on the Bentwaters incident. In describing the events of December 26–28, 1980, Jim Penniston recounted the designs he saw on the craft-of-unknown-origin and described his experience of feeling compelled to write down ones and zeros shortly after touching the craft."

Penniston says:

It was around that time that there was a pause in shooting and John asked about a certain date and I said, "Yes, John, I have that written in my notebook here." So when I started to thumb through the note-book looking for the date he asked for, I started to flip the ones and zeros pages. I was stopped by the producer and got the immediate attention of John and Linda, asking what was that written in the note-book. Casually I tell them, "Oh, it is the ones and zeros from that night." Linda jumps up and said, "You wrote them down?" and I said, hesitantly, "Yes." Everyone was excited with what I have felt for thirty years was a product of my temporary madness from exposure to the craft that night. No one knew there at the film shoot that I was reluc-tant and also ashamed of what the ones and zeros meant to me that morning of the next day after the incident. It was a secret I kept only to myself until filming the shoot and until I made the slip about the ones and zeros. I told Linda and the rest of the film crew that I had written these down in my notebook. The original notebook I had from that night was in my hand. Well, everyone had pretty much stopped and got excited and was curious about what I had told them. Everyone present, including John, asked me, "You recorded them?" and I said, "Yes." All eyes were on them at that point. Linda wanted some experts to see them. I agreed for her to do that . . . but unbeknown to her, I also gave copies of the codes to Kim Sheerin, for their expert to decipher.

Previously, only the first five pages of Penniston's notebook have been made available. But for the first time in this book, all sixteen pages are be-ing revealed (the raw data are reproduced as appendix B). Following on from this, the following translation is offered by Penniston, based on the work of the various experts listed earlier. It consists of brief snatches of text, together with numbers, interpreted as latitudes and longitudes, the locations of which are given in brackets:

EXPLORATION OF HUMANITY 666 8100

52.0942532N 13.131269W (Hy Brasil)

CONTINUOUS FOR PLANETARY ADVAN???

FOURTH COODINATE CONTINUOT UQS CbPR BEFORE

26.763177N 89.117768W (Caracol, Belize)

34.800272N 111.843567W (Sedona, Arizona)

29.977836N 31.131649E (Great Pyramid in Giza, Egypt)

14.701505S 75.167043W (Nazca Lines in Peru)

36.256845N 117.100632E (Tai Shan Qu, China)

37.110195N 25.372281E (Portal at Temple of Apollo in Naxos, Greece)

EYES OF YOUR EYES

ORIGIN 52.0942532N 13.131269W (Hy Brasil)

ORIGIN YEAR 8100

Here are Penniston's final thoughts on the binary code and on the possibility that this is a message from the future. It is a bizarre mixture. At times Penniston seems certain about things, but elsewhere it's clear that he's plagued by anxiety and self-doubt:

So the above is the binary code decipher, or as I refer to it the binary enigma. As I have always said from the very beginning of my exposure and witnessing of the incident, it was clear that I could with 100 percent certainty tell you what the craft was not. The hard part is this: what exactly was it? It was not an airplane in the Jane's Book of Known Aircraft, or many other things. What I cannot tell you is exactly what it was. It is just like this binary? How can it be possible to receive such information from contact that night? How could I go home and 24 hours later write these ones and zeros down from memory? How, why, and a thousand more questions I have. Under hypnosis, I reveal that they are time travelers from the future.

I have a detailed tale of them while under this hypnosis in extreme detail. Then I have the real evidence: the written binary, the day after the event; the glyphs observed on the side of the craft; the pod casts that were made later on the morning after the event. Why does this translate in English and not some alien language that cannot be understood? Why did the aircraft have remotely terrestrial similarities?

So how can it be that the physical evidence seems to back up the hypnosis? It is all good food for thought, I guess. Will we really have the answers to this enigma? My thoughts on this: a firm yes.

During my investigation and in the course of my research with the time traveler evidence (as it kept coming up at various points when looking for the definitive answers to Rendlesham and the enigma), the answer to that question is that there are no definitive answers at this point in time. In recent years I find that Rendlesham is not the only incident which other military have evaluated as time travelers. I find that I am not the only one who has suggested this, by looking at the evidence of a particular case. The same thing happened with Project Avalon: they were talking about the Roswell visitors and Admiral George Hoover said that "these visitors were us from the future. They were time travelers; they weren't extraterrestrials." Now other researchers and whistle-blowers have spoken about the same thing and this is fascinating in itself. But what Admiral George Hoover said "the biggest secret" really was—it had to do with the abilities and the power of the consciousness of these travelers. Because they were us from the future, what the military authorities had found out was what humans are really capable of. This had been buttoned up really tight, because if we knew how powerful we really were, how powerful we really could be, then we would cause chaos around us, and this could never be permitted. We could rearrange the reality around us in the way that we wanted to, in the way that—if this is real—the future humans had learned how to do, which gives them access to these sorts of incredible abilities, such as time traveling.

I have my own beliefs, but in no way expect you to blindly go with my thoughts or my beliefs on this. If I were walking in your shoes, well, I would look at the evidence that has been presented in this book. I would then take a step back and then would look at all the evidence and systematically evaluate the evidence with the binary and draw my final conclusions myself. You must decide is this as Jim Penniston be-

*lieves, that it is time travelers from our future came back in time to
1980, or is it something totally different than that? It is for you to decide. A question running through my thoughts is this: why did I decide
for this binary message to be made public? After all, it could be con-
ceived as a private message, for only me to know. But then I ask why
would it be me? I am only one of many, so I think that the only answer
that I can give is the most honest and simple. So my thinking is that it
is for all of mankind and not one man. After all, if they are really us
from the future, then this is a message for all of us, from us in the future, giving me this conclusion for myself about the binary: what if the
whole point to the contact in December 1980 was for us to publish this
binary message at this point in time? It is my thought, but if they are
from the future the great part of that thought is that sometime in the
future we will know if it is true or not. Is there more to come? Yes, I
believe there is much more to come!*

It is extremely difficult for me to evaluate the material in this chapter.
Indeed, I was tempted to conclude with the age-old phrase used by intelligence analysts when confronted by intriguing claims from a single
source, unsubstantiated by a secondary source: "interesting, if true." But
readers will expect something less glib and Penniston deserves better
than this.

First, is Penniston telling the truth? This is impossible to verify, though
I would make one observation on this point. The most compelling aspects
of the Rendlesham Forest incident are the events that were witnessed by
several people, such as when Burroughs and Penniston witnessed the UFO
at close range on the first night and when Halt and his men saw the UFO
fire light beams at the ground on the subsequent night; it's the corroborat-
ing evidence such as the charred trees, the radiation readings, and the radar evidence. Penniston is the first to acknowledge that the binary message
is much more problematic in evidential terms. Thus, this aspect of the
story may seem *less* credible. This being the case, Penniston has more to
lose than to gain by telling this story.

Is it possible that Penniston is confabulating, as mentioned previously
in this chapter? Confabulation (sometimes called honest lying) is where
someone is telling the truth as they perceive it but where events did not

take place as described. This can be the consequence of a brain injury, Alzheimer's disease, or any number of psychological or psychiatric disorders. Might all this be the consequence of the drugs and hypnosis that many claim were administered after the event, during the debriefings and interrogations? Penniston himself recalls only one debriefing from the AFOSI, with all others administered by the command element. Could it be a result of the regression hypnosis Penniston underwent voluntarily, with a hypnotherapist? The hypnotherapist wasn't a ufologist and Penniston had only attended because he wasn't sleeping properly, so it's unlikely that the hypnotherapist led Penniston down this road. Or could it be a consequence of a mixture of these various factors? However, the pages of Penniston's police notebook, which contain the code, are dated December 26 and 27, 1980. And Penniston has always maintained an absolute willingness to have the notebook forensically tested and dated, if this is possible, to prove the code wasn't added later. So Penniston himself is adamant that this is not the consequence of drugs and hypnosis.

Here is what Penniston wrote about this in 2010, in an attempt to make sense of it all:

Any inconsistencies in my account can easily be attributed to the meddling of the inept debriefing and the drug-induced attempted extraction of information by the U.S. agents at the AFOSI building, or quite possibly by the phenomenon itself. The other factor is simply my state of mind at the time of the incident. I was shaken deep within every fiber of myself by all that I saw and experienced, in the midst of doing my job that night. It was more than unraveling and disturbing for me. It was "life-changing." This does not take into account my private ordeal at home with the binary hell I was encountering, as well as other things immediately after the incident. I believe I was possibly a primary reason for Colonel Halt's inconsistencies in his infamous sanitized memo as well and the initial confusion about many things with the incident. Of course, then causing more of the initial confusion was the blanket of silence directed by the base command; I also think this was also a major factor.

His final words on this, written more recently, in 2013, are clear: "I have always have had memory, and my children too, of the glyphs and ones and zeros."

What of the message itself? If I still worked for the MoD, I would have secured a proper analysis from cryptographers at GCHQ (Government Communication Headquarters)—a UK agency broadly analogous to the NSA. But with this option no longer available, I can offer few personal thoughts, not being an expert in codes. However, with the greatest of respect for those experts who have worked on the message, I would take a step back and first ask if it was binary code at all. That may seem a logical assumption, but it's an anthropocentric one. If the message is genuine and has been correctly decoded, it seems simultaneously profound and banal. "Exploration of humanity. Continuous for planetary advance" is the best approximation of the textual part of the message. It sounds impressive on the face of it, yet arguably it's no different from all sorts of channeled messages circulating in the UFO and New Age communities. The same might be said of the locations. Hy Brasil is a mythical lost land said to be off the coast of Ireland. It's sometimes dubbed the Celtic Atlantis. The other sites read like a New Age holiday wish list: the Great Pyramid, the Nazca Lines, and Sedona. Again, is this proof that the message is genuine or that it is confabulated? Have the various experts somehow shoehorned the data into something that fits their own belief systems? Is all this just wishful thinking? Or is the message more subtle? Is there a more complex message hidden deeper within the obvious one? I have no answers here, which is why we provide the raw data in appendix B. Maybe other experts will come up with an alternative translation.

Finally, I should say that while I worked on the MoD's UFO project and was aware that the UK government dabbled in other paranormal areas such as remote viewing, I am not aware of any UK government research into time travel. I am aware of some rumors that the US government conducted such research, but while stories such as the Montauk Project and the Philadelphia Experiment have been widely told, I have never come across any official corroboration of this.

With the Rendlesham Code revealed (though not necessarily explained or decoded) the time has come to conclude this extraordinary story, as John Burroughs and Jim Penniston try to make sense of not just the codes but the whole incident.

In this next chapter, I make no comment at all. Burroughs and Penniston wanted it this way, so that their material would stand alone. They, after all, were the ones who experienced this and had to live with what happened next.

19. FINAL THOUGHTS FROM JOHN BURROUGHS AND JIM PENNISTON

The night we walked into Rendlesham Forest, I, Jim Penniston, went in as a Security Police supervisor, with only one thing in mind, to ascertain what exactly it was and, if I confirmed that it was an aircraft crash, render assistance to survivor(s), set up an Entry Control Point for responding emergency personnel, and, if it was ours, set up a National Defense Area around it, which would have secured the perimeter and entry for U.S. personnel only.

But that was not the case. After what I had seen in the early-morning hours on that night, I left that forest a different man than when I walked in. I was in awe of the technology and, yes, a knowing that it was not an aircraft that could have been manufactured in 1980 or even now. The replication of this craft in my assessment is far more advanced than any we could develop in the next few centuries at the very least.

I have had difficulty with what I saw and was exposed to out there that night. Within weeks after the incident I was being treated for new aliments I had not had before: vertigo, headaches, and then memory issues and unknown infections: all treatable but always reoccurring. The interesting thing was that I never thought of time out there in the forest with John immediately after the incident, as if it was only an experience that no doubt could never be explained. I really can't explain that, since the night of the incident, it all seems pushed to the back burner for me, or at least through my remaining time in the USAF. After retirement, a unique and strange thing happened. For after a void of nightmares and dreams about Rendlesham, then after my retirement it all started to happen. For the last twenty years I have also suffered from chronic fatigue syndrome. Of course, as time went on it became more prevalent. I can say it was an evolving process for me. I would look occasionally at the notebook that had the strange glyphs, notes, and the memory of my madness after the night.

I often think, so if my memory was messed with, then the obvious question is why? And if this was done by our own government, then there were things used for containment purposes, but then why would they allow the information about the binary codes and glyphs to exist in my notebook? I am sure that information could have all been extracted under drugs or in other parts of the interrogation, as described in my one and only hypnosis session. Then of course there is the recall of information from going back to the landing and takeoff area at Rendlesham. Then other recall from one of the binary locations. All of this and my question is, of course, why? The conduct of the agents, while I was under the operational control of the USAF, makes me wonder why the Air Force would allow such things to be done to their personnel while on a base! Why would they, my fellow Americans, treat us as enemy combatants? Or even worse.

Or is the answer as simple as this: When I was taken into the building for debriefing and sodium pentothal, the agents revealed nothing, for they were met with silence and what I knew would remain silenced for years. Only to be released by design. Just a thought; who really knows?

I do take medication that curtails the nightmares and flashbacks from the night. The diagnosis is PTSD. I suppose those things are just normal for me today. As I search for the answers and live with the debilitating effects from that night and follow-up days, by my countrymen's interrogations, there are one or two questions that I want to ask and have answered. They are: "What was it that we saw and why have the United States and the United Kingdom done everything to cover it up as a UFO event?" and "What is it they are so afraid of the public finding out?"

Whatever it is that they want to be kept so secret obviously has nothing to do with UFOs (ETs or aliens), as is clearly shown in the declassified documents, which clearly talk about this phenomenon. But I could not summarize this event without talking about General Williams (Wing

Commander at the time) just a little bit and some of the things he has said to me that have made me feel like we are going where John and I should be with this incident. Recently I have been in contact through e-mails and phone calls with the former wing commander, the now-retired Major General Gordon Williams. John and I have recently met with him, too. The general was and is always willing to talk and meet with us. During one of the e-mail exchanges I was having with him, I felt guilt ridden to tell the general that on the day I briefed him I was under orders of the AFOSI and they said an active investigation was underway. The information I told him and the other staff officers was the AFOSI-sanitized version of the event and did not go into all the detail of my exposure with my team with the phenomenon. The general said, "I know you guys were not treated fairly and [the AFOSI agent] could be real tough." I was thinking to myself, How interesting, since I knew [the agent General Williams named], or at least what he looked like, and he was not one of the people who interviewed me. I remember when I was visiting John in Arizona, to gain information for this book, I had left a voice mail on General Williams's cell phone, trying to arrange a meeting with General Williams, John, and me in Tucson, where the general lives. I promptly received a callback from the general while I was with John and others at the Stupa in Sedona. The general agreed to meet us in a few days. The one thing that was so strange with the general was what he said to me on the phone call setting up the meeting. He said, "You guys are the most persistent sons of bitches I have ever seen, good on ya! You're just like fighter pilots; you see your target and ya don't stop till ya get it done." I found this very characteristic of General Williams, minus the charm school stuff. He always seemed so supportive; in a strange way he seemed proud that we were continuing our quests for the truth.

The general agreed to meet with John and me on the weekend. When we met at a small café in Tucson in October 2012 it was a strange sensation as we all shook hands. Sitting down in the booth, John and I were on one side and across from me was the man whom I had not seen for thirty-two years. My mind wandered a bit in thought as I ran back to the time I faced this man with those steel blue eyes looking back at me. I realized the last time I looked this man in the face was the day I was reporting with John in his office, at the 81st Wing Commander's office, reporting what had happened on night one. To my left in a small seating area in the general's office were seated Colonel Conrad and Colonel Halt, listening to us attentively. I gave the general the brief that was prepared for me. What was so curious at the time was that when I finished with my brief to him he said, "I want to thank you men for doing a great job, and I thank you for your candor; we will take it from here." I left the office with a persistent question about that meeting: it was why the general did not ask one question of us. How strange I thought that was then. Little did I know that thirty-two years later I would learn why he didn't. It was because of the declassified document explaining the phenomenon, and he was aware at the time of it.

When John was ready to let the general know that we had found the declassified document and asked if he would like to see it the general, without emotion, said, "Don't tell me, you have a stamped document?"

John said, "Yes, sir, that is what we have." John got up and excused himself for a few minutes and it left the general and me at the booth. The general took his index finger and followed each line and read the document, which stated all wing commanders were briefed on this phenomenon and what he was supposed to do with it if it came up under his watch. Well, little did he know it would happen to him. The general said, "Whatever you need me to do, just ask." John and I left, with the general picking up the lunch tab.

What has always made me wonder the most about those officers over me was their ongoing support. Not just General Williams but also Colonel Conrad and others who were involved. It seems they are offering support not so much for us but because they have guilt involved with what seems like support. Was how all the officers over me at the time still support us done without question? Or maybe it is only that they continue to follow the orders and the program that dealt with this phenomenon from so long ago. I was also haunted with what I knew from being out in the forest that night and what I observed, keeping secrets that I can only have told at this point in time.

After John and I met again for the first time in mid-2009, the first time we really ever sat down and started discussing the Rendlesham Forest incident, new enlightenments of time loss and further human tragedy began to surface, as at our meeting we began to recall and remember more. We soon began to realize that dreams and nightmares were actually pieces of our persistent memory that were finally coming forward after years of suppression. The following is how John remembers the incident and the strange things that followed and what has brought us to this point in history.

Our journey has never really ended from the night we entered the forest until the present day. Something inside of us through the years has continued to push us toward finding out what we encountered over those three nights, December 26–28, in Rendlesham Forest. As Jim has said, "What I once believed is no more and what I've witnessed defies all that I have ever imagined. I am truly in awe over the whole incident, and no one can fully understand the magnitude of such an event, unless they were there." We were brought in after the incident and treated like enemy soldiers would be treated if they had been captured. Jim remembers being interrogated and having drugs used on him. What I remember is being called in and asked to write a statement by the Deputy Base Commander, Charles Halt. After I wrote my statement, Staff Sergeant Penniston, Airman First Class Cabansag, and I were debriefed in Halt's office. We were told not to talk about the incident to anyone, that there was an ongoing investigation, and then AIC Cabansag, Jim, and I were dismissed and taken into the Wing Commander's office with the Base and Deputy Base Commanders present. The AFOSI-sanitized account was briefed to the officers. The Wing Commander, Colonel Williams, thanked us for the report and asked no questions, which was the beginning of how our command staff left us to hang out on the clothesline to dry.

In the spring of 2009 I decided that it was time to do something about what had happened to us. I was tired of reading and hearing about what happened to us from non-witnesses. So I decided to do an investigation to try to put everything together. The first thing I did was contact the guy who stood next to me in the field that cold December night in 1980. I contacted Jim via e-mail and followed up with several phone calls, with us finally meeting in a small café on Highway 39, halfway between Freeport and Normal, Illinois. We decided to work together on this and put all the pieces together once and for all. As we drove away from that dinner in that late-summer afternoon neither one of us had any idea of the things to come! I returned to Phoenix and Jim and I started putting the pieces together. As we did so, we went back and forth on what would be the best way to get all of it out after we finished the investigation, finally deciding to put it in a book.

What we learned along the way shocked us to the core. In October

2011 I came across an article written by Nick Pope talking about Project Condign (see chapter 14). After reading the article I contacted him asking him if there was more to the story. He sent me back a report, which I read over, leading to more questions. One of the questions I asked was what files were used to write the report. He then sent me a PDF file that contained all of those documents. When I opened up that file I was blown away by the information the documents contained. This report talked about our incident. It talked about how if certain statements were made by witnesses then they knew that they had a phenomenon—a phenomenon that USAF wing commanders had been briefed on—because of the incident that had happened with military aircraft—and then a team would be sent in to investigate the incident and interview the witnesses. One year later, in October 2012, while Jim was in Phoenix, we confirmed in a café in Tucson with General Williams that he was aware of the phenomenon, that the stamped document we showed him was real, and that he was surprised it had been declassified.

As I went through the documents I also noticed that some of the information was still classified, for a variety of different reasons, so I asked Nick who the author was and whether some of this information been declassified since the files were released. His response back to me caught me off-guard. He stated he was not a whistle-blower and he could not comment further on the documents. He did say this to one of my questions about why more of what was in these documents had not been discussed: "Many ufologists have looked at the Condign Report and made FOI requests in relation to it. The problem is that it was written by someone with a background in scientific and technical intelligence and was very much a document for a government/military audience. As such, without wanting to be overly patronizing, ufologists aren't really capable of analyzing it and undertaking meaningful follow-up. There was a sort of schoolboyish excitement in the UFO community that such highly classified documents had been made public, but they didn't really understand what they'd got." So this might explain why even after those documents were released ufologists did not follow up on what they had been given.

One of the cover-ups we believe should be exposed (see chapter 13) is this, taken from a BBC News report by Neil Henderson dated March 2, 2011:

UFO FILES REVEAL "RENDLESHAM INCIDENT" PAPERS MISSING

Intelligence papers on a reported UFO sighting known as the "Rendlesham incident" have gone missing, files from the National Archives reveal.

The missing files relate to a report of mysterious lights from US servicemen at RAF Woodbridge in Suffolk in 1980.

The disappearance came to light with the release of 8,000 previously classified documents on UFOs.

Officials found a "huge" gap where defence intelligence files relating to the case should be, the papers show. . . .

These classified documents need to be released to the public.

In the past three years I have developed a serious heart issue—almost dying in July of 2012—and I now have a pacemaker defibrillator controlling my heart. My doctor needed to see my military medical records because he could find no reason why I was having this heart issue. With the help of attorney Pat Frascogna I contacted Senator John Kyl's office with a letter and support documents asking for his help. Senator Kyl's first attempt to get my records was denied by the US VA. They told him I needed to file a VA claim for disability, and then came the most stunning part of the letter; they told him they felt my records could be located in the VA classified-records section and that if they were it would be up to the USAF to release those records to the Senator and my doctor for treatment or even allow the VA to treat me.

I was shocked my country could be holding up my treatment because my medical records were classified. What it clearly shows is this: that I, along with many other veterans, can have my health put on hold while someone decides if I was worth being treated or not. I filled out the paperwork in June and turned it into the VA. I then notified the Senator's office that a claim had been filed. After hearing nothing back for several months I sent Senator Kyl another letter, mailing it on Veterans Day 2012. In December I received a response back saying he was retiring and that he was unable to get my medical records and I needed to try again with Senator John McCain's office. So a new packet has been sent to Senator McCain's office. Senator McCain has replied back saying he will see what he can uncover. Will Senator McCain be able to get information that no one else has been able to get their hands on? Information that will help Jim and me, who served our country for over a quarter of a century, over half of our lifetimes, and will allow us to get the treatment we need? Or because our records remain classified will we continue to suffer from the effects of exposure to some as yet unknown phenomenon—the existence of which was known and under study by the U.S. military—receiving no medical care, given no information, allowed to continue languishing in the silence surrounding the events of that week, with us being some of the last victims of the Cold War!

In closing, it's our belief that the Rendlesham Forest incident is a bigger and more significant UFO case than Roswell. It has been devilled by misinformation, disinformation, and people wanting to write themselves into the story. It has been our intention, in this book, to place in the public domain everything that we know about the extraordinary series of events that took place at Bentwaters and Woodbridge, both during the encounters themselves and in the aftermath. While there have been previous books on the subject, they have been written by people in the UFO community. Now the military personnel at the heart of this incident will finally have their say. We intend to set the record straight and tell the full story of these extraordinary events for the first time. We do so in order to reveal the truth about events that we believe are of immense historical significance and public interest. We also do so for the men and women stationed at Bentwaters and Woodbridge at the time, many of whom have suffered as a result of what happened. It is our hope that the publication of this book will lead to the wrongs they have suffered being righted. We had hoped to show why Nick Pope made this statement "I've gone on record saying Rendlesham might be the turning point in history that leads to the explanation of the UFO phenomenon." As far as we see it, is there more to be told? Absolutely! This book describes the events from A to Z on the historical aspects of the Rendlesham Forest incident. It goes into further detail about some of the supporting information. It does cover the binary release and some thoughts on what we, the witnesses, think. It is the most accurate and factual book written to date. But the real questions for the readers to ask are: Is this really the story of the witnesses and has everything from the witnesses been addressed? Is there more to come? Yes, there is more to come.

CONCLUSION

This, then, is the story of the Rendlesham Forest incident. It has been the story not just of the world's best-documented and most compelling UFO encounter but also of the effect these events have had on the two military men at the heart of the incident.

What have we learned about these sensational UFO sightings that really are bigger than Roswell?

We know that a UFO landed next to one of the most sensitive military installations in the NATO military alliance.

We know that the UFO was seen on three consecutive nights by dozens of highly trained military personnel and that the witnesses included the Deputy Base Commander.

We know that light beams from the UFO struck the ground just feet in front of the Deputy Base Commander and a party of men and that later the UFO was seen firing light beams onto the base and, in particular, into the Weapons Storage Area.

We know that the UFO was tracked on radar.

We know that there was physical trace evidence at the landing site, including damage and scorch marks on the trees and radiation levels that the MoD's scientific staff assessed as being significantly higher than background levels, raising the terrifying possibility that John Burroughs, Jim Penniston, and maybe other witnesses were irradiated by this craft.

We know that in a UK study classified "Secret UK Eyes Only" intelligence staff expressed the view that "several observers were probably exposed to UAP radiation," when the US government will not even acknowledge that the incident occurred.

We know that despite the US government stating that UFO sightings have not been investigated since 1969, not only was this incident investigated, but also the senior USAF general in Europe flew in to be briefed and removed evidence, without telling the UK government.

We know that a former Chief of the Defence Staff in the United Kingdom, a five-star military officer, believed his own government was covering up the truth about this incident.

We know that some of the key files and documents that might have provided answers about what happened have apparently been destroyed or lost in mysterious circumstances.

We know that military medical records for John Burroughs and Jim Penniston are lodged in a classified-records section of the Department of Veterans Affairs and that they have not yet been released, despite extensive and high-level lobbying and the threat of legal action.

We know that while the US and UK governments have consistently sought to downplay or ridicule the UFO phenomenon, behind the scenes the subject is taken extremely seriously but may largely have been moved into the private sector to take it outside the scope of FOIA legislation and to make congressional scrutiny more difficult. The interest would be further hidden by avoiding using the phrase "UFO."

What we *don't* know, with a lot of this, is *why*. Why the secrecy? Why can't the witnesses be briefed on what happened to them? Why can't the public be told the truth about the Rendlesham Forest incident and about some of the other cases mentioned in this book? There are some intriguing clues. Project Condign talked about the possibility of "weaponization." If there *is* an unknown technology here—whatever the source—the nation (or corporation) that first acquires it and masters it will be in possession of something of incalculable power and value. This, we suspect, means the stakes are sufficiently high in some people's minds to justify the secrecy, the lies, and the hanging out to dry of the loyal military personnel who were caught up in all this. The only alternative is that the secrecy goes beyond technology acquisition alone and has been put in place to prevent some revelation so earth-shattering that the powers that be would go to almost any length to prevent disclosure. What might that secret be and could the Rendlesham Code be the key to this? Only time will tell.

Despite the obstacles being put in their path, John Burroughs and Jim Penniston continue to press the authorities for answers and maybe an apology. These events have had a profound (and overwhelmingly negative) effect on their lives, not least in terms of ongoing health issues. Their campaign, spearheaded by their tenacious attorney, Pat Frascogna, will continue. It is hoped that this book will play a key part in raising the profile of these events and, ultimately, in providing the explanation—and thus the closure—that Burroughs and Penniston want.

While Burroughs and Penniston were the two witnesses most closely involved in the Rendlesham Forest incident and have been the focus of this book, dozens of other military personnel stationed at the twin bases of Bentwaters and Woodbridge witnessed the same craft Burroughs and Penniston encountered or were involved in the events in some other way. Some of these stories have been told in this book, but other witnesses have yet to come forward. Burroughs and Penniston are campaigning not just for themselves but for everyone involved.

It is hard to escape the inevitable conclusion that someone in the US government, military, or intelligence community knows more about this than they are currently saying. This may be out of embarrassment. After all, the US government has consistently told the media and the public that it has no interest in UFOs and has not investigated sightings since Project Blue Book was closed down at the end of 1969. The witnesses and the USAF's own documents tell a very different story. But it goes much further than this. We know from a declassified MoD document that the senior USAF officer in Europe visited the bases in the immediate aftermath of the incident, was briefed on it, and—to the absolute fury of the MoD— removed evidence without briefing the UK authorities. That's where the trail goes cold, but clearly somebody knows what happened next.

The prediction that Burroughs and Penniston made at the end of the final chapter, that "there is more to come," has already come true. From April 29 to May 3, 2013, an event called the Citizen Hearing on Disclosure was held at the National Press Club in Washington, D.C. Organized by Stephen Bassett, a registered lobbyist who has long campaigned to end what he calls the "truth embargo" about UFOs, the format of the event mirrored the congressional hearings on UFOs that many in the UFO community want to see.

This was certainly no conventional UFO conference. The involvement of a number of former Congressional Representatives (Republican, Democrat, and Libertarian) brought genuine gravitas and discipline to the proceedings. The researchers and witnesses were arranged onto themed panels and gave evidence to the committee of former Representatives. Each panelist read an opening statement on to the record, but thereafter the questions were at the discretion of the committee.

Unsurprisingly, John Burroughs and Jim Penniston were the ones who most caught the attention of the committee. Burroughs and Penniston discussed their concern that they were irradiated by their close proximity to whatever it was that they and so many other military witnesses encountered back in 1980. They expressed their frustration with their inability to secure the release of their medical records from the classified-records section of the VA, and Pat Frascogna briefed the committee on the perceived stonewalling in relation to the FOIA requests and then announced the intention to file a federal lawsuit.

The politicians on the committee were so outraged by the treatment meted out to Burroughs and Penniston that they contacted them immediately afterwards, both directly and through staffers, and signed their names to a letter requesting that the medical records of Burroughs and Penniston be released, so that they can receive a proper diagnosis and treatment.

The letter was sent to the Secretary of Veterans Affairs and to the President of the United States. Perhaps, finally, this will help unlock the secrets of what happened in Rendlesham Forest and help provide some answers.

That's all Burroughs, Penniston, and the other military witnesses want—answers. After all these years and after the loyal service they've given their country, they deserve this much.

SELECTED DOCUMENTS AND IMAGES

APPENDIX A: LIEUTENANT COLONEL CHARLES HALT'S OFFICIAL REPORT OF THE UFO ENCOUNTER

DEPARTMENT OF THE AIR FORCE
HEADQUARTERS 81ST COMBAT SUPPORT GROUP (USAFE)
APO NEW YORK 09755

REPLY TO
ATTN OF. CD

13 Jan 81

SUBJECT: Unexplained Lights

TO: RAF/CC

1. Early in the morning of 27 Dec 80 (approximately 0300L), two USAF security police patrolmen saw unusual lights outside the back gate at RAF Woodbridge. Thinking an aircraft might have crashed or been forced down, they called for permission to go outside the gate to investigate. The on-duty flight chief responded and allowed three patrolmen to proceed on foot. The individuals reported seeing a strange glowing object in the forest. The object was described as being metalic in appearance and triangular in shape, approximately two to three meters across the base and approximately two meters high. It illuminated the entire forest with a white light. The object itself had a pulsing red light on top and a bank(s) of blue lights underneath. The object was hovering or on legs. As the patrolmen approached the object, it maneuvered through the trees and disappeared. At this time the animals on a nearby farm went into a frenzy. The object was briefly sighted approximately an hour later near the back gate.

2. The next day, three depressions 1 1/2" deep and 7" in diameter were found where the object had been sighted on the ground. The following night (29 Dec 80) the area was checked for radiation. Beta/gamma readings of 0.1 milliroentgens were recorded with peak readings in the three depressions and near the center of the triangle formed by the depressions. A nearby tree had moderate (.05-.07) readings on the side of the tree toward the depressions.

3. Later in the night a red sun-like light was seen through the trees. It moved about and pulsed. At one point it appeared to throw off glowing particles and then broke into five separate white objects and then disappeared. Immediately thereafter, three star-like objects were noticed in the sky, two objects to the north and one to the south, all of which were about 10° off the horizon. The objects moved rapidly in sharp angular movements and displayed red, green and blue lights. The objects to the north appeared to be elliptical through an 8-12 power lens. They then turned to full circles. The objects to the north remained in the sky for an hour or more. The object to the south was visible for two or three hours and beamed down a stream of light from time to time. Numerous individuals, including the undersigned, witnessed the activities in paragraphs 2 and 3.

CHARLES I. HALT, Lt Col, USAF
Deputy Base Commander

APPENDIX A: LIEUTENANT COLONEL CHARLES HALT'S OFFICIAL REPORT OF THE UFO ENCOUNTER

APPENDIX B: JIM PENNISTON'S "BINARY CODE MESSAGE" FROM HIS ORIGINAL POLICE NOTEBOOK

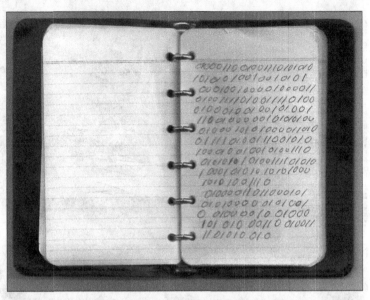

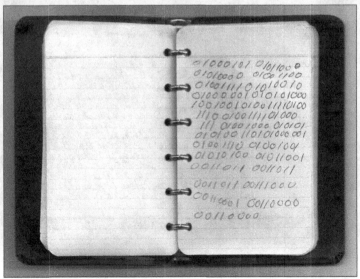

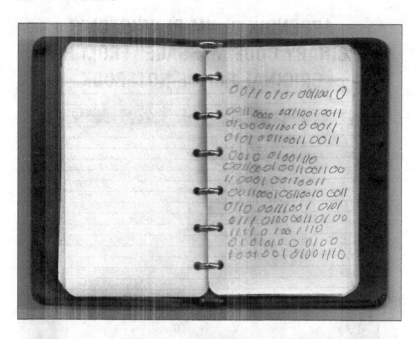

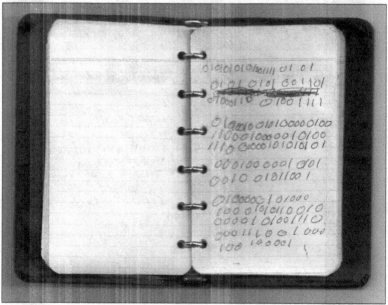

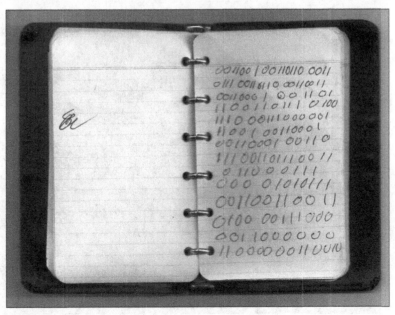

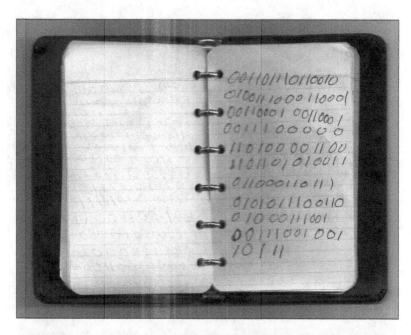

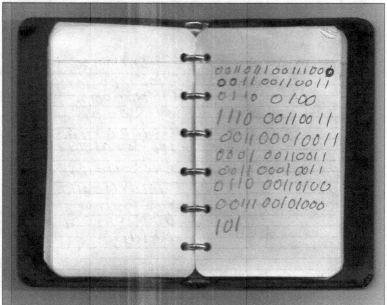

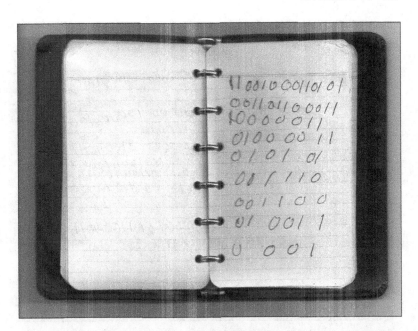

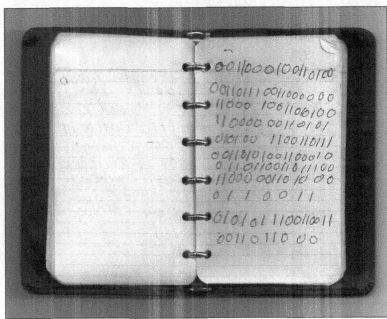

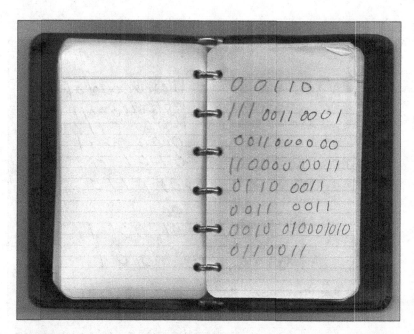

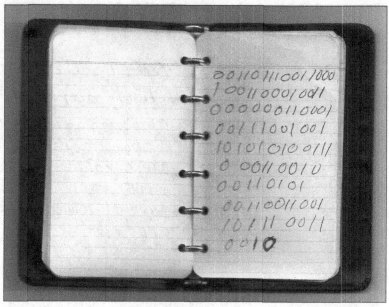

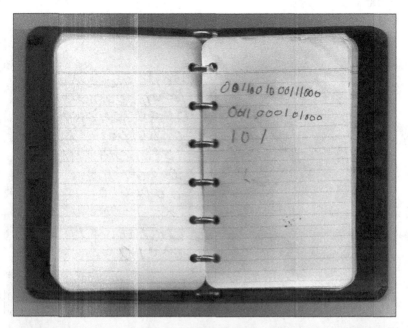

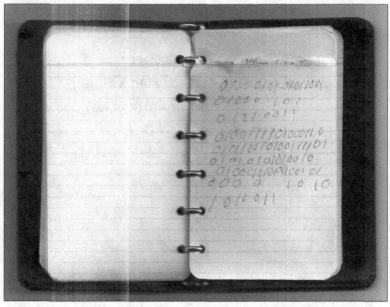

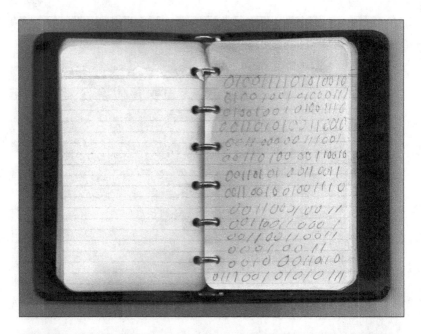

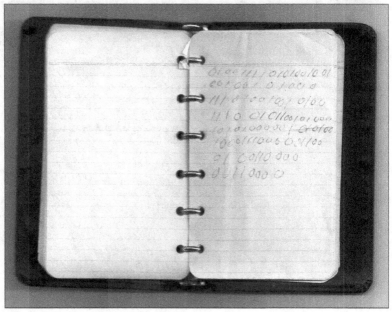

APPENDIX C: A DEFENCE INTELLIGENCE STAFF DOCUMENT ASSESSING THE RADIATION READINGS TAKEN AT THE LANDING SITE

E13

Loose Minute

DI52/103/10

OR 24/2

DI55a Attn Mr C P Cooper

Unexplained Lights

Reference: DI55/103/15/1

1. Like DI55, DI52 do not know of any serious explanation for the phenomena described at reference.

2. Background radioactivity varies considerably due to a number of factors. The value of 0.1 milliroentgens (mr), I assume that this is per hour, seems significantly higher than the average background of about 0.015 mr. I would not expect the variation in this to be much more than a factor of two, although it might be greater for specific reasons.

3. If you wish to pursue this further I could make enquiries as to natural background levels in the area. The way the US report is written, however, suggests that 0.1 mr was greater than they expected.

R C Horcroft
ADI/DI52

23 February 1981

APPENDIX D: A DEFENCE INTELLIGENCE STAFF DOCUMENT CONFIRMING THAT CINCUSAFE VISITED THE BENTWATERS/ WOODBRIDGE BASES IMMEDIATELY AFTER THE UFO ENCOUNTER

LOOSE MINUTE

D/DD Ops(GE)/10/8

DS8

UNEXPLAINED LIGHTS

Reference: A. D/DS8/72/1/2 dated 20 Jan 81.

1. At Reference you forwarded a report from RAF Bentwaters for information and asked if anyone else might have an interest in the content. You will see from the attached TM, I forwarded a copy to DI55 and PS/ACS(G)(RAF). I have had no response.

2. SOC/CRC Neatishead regret that the radar camera recorder was switched off at 1527Z on 29 Dec 80 and an examination of the executive logs revealed no entry in respect of unusual radar returns or other unusual occurrences.

3. I have spoken with Sqn Ldr Moreland at Bentwaters and he consider the Deputy Base Commander a sound source. I asked if the incident had been reported on the USAF net and I was advised that tape recorders of the evidence had been handed to Gen Gabriel who happened to be visiting the station. Perhaps it would be reasonable to ask if we could have tape recordings as well.

16 Feb 81

D BADCOCK
Sqn Ldr
Ops(GE)2b(RAF)
MB 4258 7274 MB

APPENDIX D: A DEFENCE INTELLIGENCE STAFF DOCUMENT CONFIRMING THAT GROUSAFE VISITED THE SENTWATERS/WOODRIDGE BASES IMMEDIATELY AFTER THE UFO ENCOUNTER

APPENDIX E: TRANSCRIPT OF THE TAPE RECORDING LIEUTENANT COLONEL CHARLES HALT MADE DURING HIS ENCOUNTER

Total Time: 18:13:00

00:00.000

Halt: One hundred fifty feet or more from the initial, [background voices] I should say suspected, impact point. Having a little difficulty, we can't get the light-all to work.

Halt (checking on recording at later date): Does it work?

Halt: There seems to be some kind of mechanical problem. Let's send back and get another light-all. Meantime, we're gonna take some readings from the Geiger counter and, er, chase around the area a little bit waiting for another light-all to come back in.

Intrabase radio transmission—Bustinza: Six . . . standing up.

Intrabase radio transmission: Bustinza to Security Control.

Intrabase radio transmission—Bustinza: Yes?

Englund: . . . that's mark one of the pod . . . pod number . . .

Halt: OK, we're now approaching the area within about twenty-five to thirty feet. What kind of readings are we getting? Anything?

Nevels: Just minor clicks.

Halt: Minor clicks.

Halt (during tape-over): Are you gonna feel much better after this? . . . No? [Unintelligible]

Halt: Where are the impressions?

Englund: There's a mark. . . .

Halt: Is that all the bigger they are?

Englund: Well, there's one more well defined over here.

Intrabase radio transmission: Sergeant Bustinza to Security Control.

Halt: We're still getting clicks.

Nevels: Getting clicks . . .

Intrabase radio transmission: Sergeant Bustinza . . . Bobby Diaz—Security six boarding . . . East Gate.

Halt: Can we read that on the scale?
01:36.799
Nevels: Yes, sir. We're now on the five-tenths scale, and we're reading about, uh, third, fourth increment.

Halt: OK, we're still comfortably safe here?

Intrabase radio Transmission—Bustinza: Do you all have a light-all? Full gas ready?

Intrabase radio transmission: Come in, Security. . . .

Intrabase radio transmission—Bustinza: Can you have 'em start loading . . . Security six have a light-all with gas . . . uh, please. . . .

Halt:. . . still minor readings the second pod indentation . . .

Intrabase radio transmission: Security, Base Security.

Nevels: Nope.

Halt: This one's dead. Let's go over to the third one over here.

Background intrabase radio transmission: Sergeant Bustinza . . .
[Walking noise]

Background intrabase radio transmission: Sergeant Bustinza . . .
[Geiger counter clicks]

Nevels: Yes, now I'm getting some residual.

Background intrabase radio transmission—Bustinza: Security.

Halt: I can read now. The meter's definitely . . . giving a little . . .
pulse.

Background intrabase radio transmission: [garbled] . . . copy.

Englund: About the center . . .

Halt: I was gonna say let's go to the center of the area next and
see what kind of a reading we get out there.

Englund: OK.

Halt: You're reading the clicks; I can't hear the clicks.

Englund: Yes, sir.

Halt: That about the center, Bruce?

Englund: Yes.

Halt: OK, let's go to the center.

Nevels: Yes, I'm getting more. . . .
02:01.481
Halt: That's the best deflecting . . . best deflection of the needle I've seen yet. OK, can you give me estimation? We're on the point five scale . . . we're getting . . . have trouble reading . . .

Englund: At Carbine at approximately oh-one-point-two-five hours . . .

Tape-over: cough or possible sound of piano bench

Nevels: We're getting right at, uh, a half of a milliroentgen.

Tape-over: piano music

Tape-over—Halt: Test point.

Halt: I haven't seen it go any higher.

Nevels: It's still flying around.

Halt: OK, we'll go out toward the . . .

Nevels: Now it's picking up.

Halt: This is out toward the number one indentation where we first got the strongest reading. Yeah, it's similar to what we got in the center.

Nevels: Right in the pod—it's mainly in the center.

Englund: This looks like an area here possibly that could be a blast . . . [Heavy breath] . . . it's in the center of the triangle. . . .
03:01:569
Halt: It's hard to tell. . . . Here, take this, my fingers are about to freeze.

Nevels: Yours, too.

Englund: Up towards seven- . . .

Halt: What?

Englund: Just jumped up towards seven-tenths.

Halt: Seven-tenths? Right there in the center?

Englund: Uh-huh.

Halt: We found a small blast—what looks like a blasted or scuffed-up area here. We're getting very positive readings. Let's see, is that near the center?

Englund: Yes, it is. This is what we would assume would be the dead center.

Nevels: Picking up more as you go along—the whole area here now . . .

[Clicks]

Halt: Up to seven-tenths? Or seven—

Background voice calling out: Five-fifty-five carbines!

Halt: —units, let's call it, on the point five scale. OK, why don't we do this: why don't we make a sweep—here, I've got my gloves on now—let's make a sweep out around the whole area about ten foot out, make a perimeter run around it, starting right back here at the corner—back at the same first corner where we came in, let's go right back here. . . . [Heavy breathing] . . . I'm gonna have to depend on you counting the clicks.

04:04:035

Nevels: Right.

Halt: OK, let's . . . ?

Nevels: I'll tell you as it gets louder. . . .

Halt: . . . then I can put the light on it and sweep around it.

Halt (talking loudly to someone in distance): Are we flagged?

Halt: Put it on the ground every once in a while.

Background intrabase radio transmission (in distance—garbled): . . . on by . . . by GBI.

Englund: This looks like an abrasion on the tree. . . .

Halt: OK, we'll catch that on the way back; let's go around. Move back. Hit it, there.

Englund: We're getting interest right over here. It looks like an abrasion pointing into the center of the . . .

Halt: It is.

Englund: . . . landing area.

Halt: It may be old, though—there may be sap marks or something on it. Let's go on back around. All right?

Nevels: Yeah, sir, you have to give some extension on it. . . .

Halt: Hey, this is an awkward thing to use, isn't it?

Nevels: I usually carry it about my ears, but this one broke. . . .

05:01:306

Halt: Are we getting anything further? I'm going to shut the recorder off until we find something.

Nevels: Picking up good.

Halt: Picking up? What are we up to? We're up to two, three units deflection; you're getting anything close to one pod.

Nevels: Picking up some here . . . picking up.

Halt: OK, it's still not going above three or four units.

Nevels: Picking up more, though—more frequent.

Halt: Yes.

Nevels: Closer together.

Halt: You're staying steady up around two to three to four units now.

Englund: Colonel Halt?

Halt: Yeah!

Englund: Each one of these trees that face into the blast, what we assume is the landing site, all have an abrasion facing in the same direction, towards the center, the same . . .

Halt: That's interesting. Well, let's go the rest of the way around the circle here. Turn it back down here.

Nevels: Picking up something.

Englund: Collection.

Halt: Let me see that.

Englund: You want one?

06:01:277

Halt: That's got a funny . . . that's . . . you're right about the abrasion! I've never seen a tree that's, ah . . .

Nevels: Small sap marks . . .

Halt: Never seen a pine tree that's been damaged react that fast.

Nevels: You got a bottle to put that in?

Halt: You got a sample bottle?

Englund: Put in the soil . . . put the cap on.

Halt: Here, sit this on the ground.

Ball: If you notice they're all at the same height.

Halt: From now on, let's, let's identify that as point number one—that stake there. So y'all know where it is if we have to sketch it. You got that, Sergeant Nevels?

Nevels: Yes, sir.

Halt: OK.

Nevels: Closest to the Woodbridge.

Halt: Closest to the Woodbridge base.

Nevels: Be point one?

Halt: Be point one.

Intrabase radio transmission: Screeching noise

Halt: Let's go clockwise from there.

Ball: Here you go.

Nevels: Point two.

Halt: Point two.

Ball: Go ahead, Ed.

Halt: Point two. So this tree is between point two and point three.

Intrabase radio transmission: You have Airman Burroughs and two other personnel requesting to ride 'em over on a Jeep at your location.

Ball: Tell them negative at this time. We'll tell them when they can come out here. We don't want them out here right now.

Halt: The sample . . . you're going to mark this sample number one.

Nevels: Yes, sir.

Halt: Have them cut it off, and include some of that sap and all . . . is between indentation two and three on a pine tree about . . . uh . . . five feet away, about three and a half feet off the ground.
07:06.685
Nevels: Just put it in here. . . . I've got some more . . . we'll . . .

Halt: There's a round abrasion on the tree about, ah, three and a half, four inches in diameter . . . it looks like it might be old, but, uh . . . strange; there's a crystalline—[background radio traffic]—pine sap that has come out that fast. You say there was other trees here that are damaged in a similar fashion?

Englund: Yes, all facing in toward the center of the landing site . . .

Halt: OK, why don't you take a picture of that and remember your picture; we ain't gonna be writing this down. Oh, it's gonna be on the tape.

Englund: You got a tape measure with you?

Halt: This is the picture; your first picture will be at the first tree, the one between . . . mark two and three. Meantime I'm gonna look at a couple other of these trees over here.

Nevels: We are getting some . . .

Halt: You're getting readings on the tree you're taking samples from on the side facing the suspected landing site?

Englund: Four clicks max.

Halt: Up to four. Interesting. That's right where you're taking the sample now.
08:00:967
Unknown: Four.

Halt: That's the strongest point on the tree?

Nevels: Yes, sir. And if you come to the back . . . there's no clicks whatsoever.

Halt: No clicks at all on the back—it's all on the . . .

Nevels: Maybe one or two . . .

Halt:. . . side facing the . . . interesting. The indentations look like something twisted as it dropped . . . as, you know, as it sat down on them. Looks like someone took something and sat it down and twisted it from side to side.

Englund: Uh-huh.

Halt: Very strange.

Ball?: We're looking at the same tree we took the sample off with this—what do you call it—star scope?

Englund: Uh-huh, starlight scope.

Halt: And you're getting . . . getting a definite heat reflection off the tree about, what, three to four feet off the ground?

Englund: Yes, where the spot is . . .

Halt: Same place where the spot is, we're getting a heat . . .

Ball: . . . and a spot on the tree directly behind us I picked up the same thing, and one off to your right.

Englund: All right, let me . . .

Halt: Three trees in the area, immediately adjacent to the site, within ten feet of the suspected landing site, we're picking up heat reflection off the trees.

Englund: Give me the light, Bob.

Halt: What's that again?

Englund: Well, shine the light on it again, Bob.

Halt: What, are you having trouble hearing him?

Englund: Yeah . . . keep it right on the spot, and then when you want him.

9:00:791

Halt: OK, turn the light off.

Englund:. . . You'll notice the white . . .

[High-pitched electronic whine]

Halt: Hey! You're right. There's a little white streak on the tree!

Englund: Indicates a heat . . .

Halt: Let me turn around and look at this tree over here now. Just a second—watch, 'cause you're right in front of the tree. I can see it. OK, give me a little side lighting so I can find the tree. OK. Ah . . .

Intrabase radio transmission: Alpha Two Security.

Halt: Now I've lost the tree.

Intrabase radio transmission: Security . . .

Halt: OK. Stop, stop. Light off! Hey, this is eerie.

Ball: Why don't you do the pod spots and then the center?

Halt: This is strange. Here, someone wanna look at the spots on the ground? Whoops, watch you [horn] don't step . . . you're walking all over 'em. [Clears throat.]

Englund: Pardon me, sir.

Nevels: Wanna step back?

Halt: OK, let's step back and not walk all over 'em.
10:01:011
    Unknowns talking in background

Halt: Come back here and somebody put a beam on 'em. You're got to be back ten or fifteen feet. You see it?

Unknown: Yeah.

Ball: OK, fine.

Halt: OK, lights off.

Unknown male—southern accent tape-over: He took this long to doc[ument]—

Halt: What do you think about the spot? [Long pause].

Intrabase radio transmission: Any problems? [Pause].

Unknown: One.

Halt: Yeah . . . [pause] . . . that looks the first spot? OK, that's what we call spot number three. Let's go to the back corner and get spot number one.

Unknown: [Unintelligible.]

Halt: Spot number one, here's spot number one right here! Spot number one, right here . . . do you need some light? There it is, right there. You focused?

Ball: Focused.

Halt: OK. Looking now at spot number one through the . . . starlight scope.

Ball: Picking up a slight increase in light as I go over it.
11:02.065
Halt: Slight increase in light at spot number one. Let's go look at spot number two. Spot number two is right over here. Right here—see it?

Ball:. . . That is spot one . . . [unintelligible].

Halt: OK, get focused on it. Tell me when. OK, lights off, let's see what we get on it.

Ball: Slight increase?

Halt: Just a slight increase?

Ball: Try the center.

Halt: The center spot . . . well, it really isn't the center; it's slightly off-center. It's right there.

Ball: Right here.

Englund: Let's mark that as the center.

Halt: OK . . . We're gonna get your reading on it right there.

Ball: OK.

Halt: Tell me when you're ready.

Ball: Ready.

Halt: OK, lights out. It's the center spot we're looking at now, or almost the center.

Ball: There's a slight increase.

Halt: Slight increase there? This is slightly off-center toward the, uh, one-two side. It's some type of an abrasion or something in the ground where the pine needles are all pushed back where we get a high radioac . . . ah, a high reading, about, ah . . . deflection of, ah, two to three, maybe four, depending on the point of it.

12:11:152

Ball: Someone wanna check it?

Englund: Yes.

Halt: You say there is a positive aftereffect?

Nevels: Yes, there is, definitely. That's on the center spot. There is an aftereffect.

Intrabase radio transmission: [Garbled.]

Nevels: What does that mean?

Englund: It means that when the lights are turned off, once we are focused in and allow time for the eyes to adjust—we are getting an indication of a heat source coming out of that center spot, as, ef, which will show up on—

Halt: Heat or some form of energy. It's hardly heat at this stage of the game.

Englund: And it is still . . .

Halt: Looking directly overhead one can see an opening in the trees, plus some freshly broken pine branches on the ground underneath. Looks like some of them came off about fifteen to twenty feet up—some smaller branches about an inch or less in diameter.

Halt: Oh-one-point-forty-eight. We're hearing very strange sounds out of the farmer's barnyard animals.

Englund: Forty-eight-point-seven.

Halt: They're very, very active, making an awful lot of noise.
13:03.224
Englund: It wasn't a figmentation.

Halt: You just saw a light? Where? Wait a minute. Slow down. Where?

Englund: Right on this position here. Straight ahead, in between the trees—there it is again.

Unknown: (faint) Wow.

Halt: What!

Englund: Watch—straight ahead, off my flashlight, there now, sir. There it is.

Halt: Hey, I see it, too. What is it?

Englund: We don't know, sir.

Nevels: So can I get some of those . . . more lights?

Halt: It's a strange, small red light, looks to be out maybe a quarter to a half mile, maybe further out. I'm gonna switch off.

Halt: The light is gone now. It was approximately one hundred twenty degrees from the site.

Englund: Yeah, it's back again.

Halt: Is it back again?

Englund: Yes, sir.

Halt: Well, douse the flashlights then. Let's go back to the edge of the clearing so we can get a better look at it. See if you can get the star scope on it. The light's still there and all the barnyard animals have gotten quiet now.

Englund: Yeah.

Halt: We're heading about one hundred ten to one hundred twenty degrees from the site out through to the clearing now, still getting a reading on the meter, about two clicks.

Halt: Needle's jumped, three to four clicks, getting stronger.

Englund?: Now it's stopped.
14:00.251
Englund?: Now it's coming up. . . . Hold on. There we go . . . about approximately four foot off the ground, at a compass heading of one hundred ten degrees.

Halt: All right, he's turned the meter off. Gotta say that again. About four feet off the ground, about one hundred ten degrees, getting a reading of about four clicks?

Unknown: Yes, sir. [Sneezes]. Excuse me. Now it's dying.

Halt: Now it's dying. I think it's something other than the ground.

I think it's something that's . . . something—something variable.

Englund: In that tree right over . . .

Halt: We've just bumped into the first night bird we've seen. We're about one hundred fifty or two hundred yards from the site.

Unknown: Another one . . .

Halt: Everything else is just deathly calm. There is no doubt about it—there's some type of strange flashing red light ahead.

Englund: Sir, it's yellow.

Halt: I saw a yellow tinge in it, too. Weird! It appears to be maybe moving a little bit this way? It's brighter than it has been.

Unknown: Yellow.

Halt: It's coming this way. It is definitely coming this way.

Unknown: Pieces of it shooting off . . .

Halt: Pieces of it are shooting off.

Unknown: At eleven o'clock.

Halt: There is no doubt about it. This is weird!

Unknown: To the left!
15:00.788
Halt: Definitely moving.

Nevels: Two lights—one light just behind and one light to the left.

Halt: Keep your flashlights off. There's something very, very strange. Get the headset on; see if it gets any stronger.

Nevels: I have.

Halt: OK. Give us your . . .

Nevels: Make a notation that this is on a beta reading, too.

Halt: It's on a beta reading?

Nevels: The beta shield has been removed.

Halt: OK. Pieces are falling off it again!

Englund: Sir, it just moved to the right.

Halt: Yeah!

Englund: Just off to the right.

Halt: Strange! Huh! One again left? Let's approach to the edge of the woods up there. Do you wanna do it without lights? Let's do it carefully. Come on.

Halt: OK, we're looking at the thing; we're probably about two to three hundred yards away. It looks like an eye winking at you. Still moving from side to side, and when you put the star scope on it, it sort of like has a hollow center, a dark center; it's . . .

Englund: . . . like a pupil.

Halt: Yeah, like a pupil of an eye looking at you, winking. And the flash is so bright to the star scope that it almost burns your eye.

Intrabase radio transmission (garbled): They're stopping here.

Halt: We've passed the farmer's house and are crossing the next field and now we have multiple sightings of up to five lights with a similar shape and all, but they seem to be steady now rather than a pulsating or glow with a red flash.
16:06.728
Halt: We've just crossed the creek.

Unknown: Here we go.

Halt: And we're getting what kind of readings now?

Unknown: Three clicks.

Halt: We're getting three good clicks on the meter and we're seeing strange lights in the sky.

Halt: At two forty-four we're at the far side of the farmer's . . . the second farmer's field and made sighting again about one hundred ten degrees. This looks like it's clear out to the coast. It's right on the horizon. Moves about a bit and flashes from time to time. Still steady or red in color. Also after negative readings in the center of the field we're picking up slight readings, uh, four or five clicks now on the meter.

Halt: Three-oh-five. We see strange, uh, strobe-like flashes to the . . . rather sporadic, but there's definitely something . . . uh, some kind of phenomenon. . . . Three-oh-five. At about ten degrees, horizon, directly north, we've got two strange objects, ah, half-moon shape, dancing about, with colored lights on 'em. But, I guess to be about five to ten miles out, maybe less. The half-moons have now turned into full circles as though there was an ellipse—eclipse or something there for a minute or two.

17:16.974

Halt: Now three fifteen. Now we've got an object about ten degrees directly south.

Unknown: Wait a minute to the left.

Halt: Ten degrees off the horizon.

Nevels: To the left.

Halt: And the ones to the north are moving—one's moving away from us.

Unknown: Moving.

Nevels: Moving out fast.

Unknown: This one on the right's heading away, too!

Halt: They're both heading north. Hey, here he comes from the south—he's coming toward us now!

Unknown: Weird.

Halt: Now we're observing what appears to be a beam coming down to the ground!

Unknown: Colors!

Halt: This is unreal! [Incredulous laugh.]

Halt: Three thirty, oh-three-thirty, and the objects are still in the sky, although the one to the south looks like it's losing a little bit of altitude. We're turning around and heading back toward, uh, the base.

Halt: The object to the south—the object to the south is still beaming down lights to the ground.

18:01.347

Halt: Oh-four-hundred hours. One object still hovering over Woodbridge Base at about five to ten degrees off the horizon, still moving erratic and similar lights and beaming down as earlier.

18:13.080 *END OF CASSETTE TAPE RECORDING*

APPENDIX F

A LETTER DRAFTED BY PAT FRASCOGNA, ESQ., TO THE PRESIDENT OF THE UNITED STATES, AT THE REQUEST OF AND SIGNED BY SIX FORMER CONGRESSIONAL REPRESENTATIVES, ASKING THAT THE MEDICAL RECORDS OF JOHN BURROUGHS AND JIM PENNISTON BE RETRIEVED FROM THE CLASSIFIED-RECORDS SECTION OF THE DEPARTMENT OF VETERANS AFFAIRS AND RELEASED TO THEM

May 3, 2013

President Barack Obama
The White House
Washington, D.C. 20500

Dear Mr. President:

As former Members of Congress, we are writing to express concern over the lack of response by our government to two of our retired servicemen who dutifully served their country during the Cold War, Air Force Staff Sergeant James William Penniston and Airman First Class John Frederick Burroughs.

In 1980 Sergeant Penniston and Airman Burroughs were stationed at the twin base facility of RAF Bentwaters and RAF Woodbridge near Suffolk, England. On December 26, 1980 these men were the first to respond to an incident just beyond the East gate of the Woodbridge base. As a consequence of their arrival on scene of that incident they almost immediately thereafter began suffering adverse health effects. These health issues have plagued them to the present day and have, at times, been life-threatening.

Despite repeated attempts to obtain their medical records from the Department of Veterans Affairs they have been denied access to them. We are respectfully asking for your assistance in making certain that Sergeant Penniston and Airman Burroughs be provided their service medical records commencing from their respective dates of induction into the United States Air Force.

Thank you for your consideration of our request.

Sincerely,

Merrill Cook
Former Member of Congress, Utah

Carolyn Kilpatrick
Former Member of Congress, Michigan

Roscoe Bartlett
Former Member of Congress, Maryland

Mike Gravel
Former Member of Congress, Alaska

Darlene Hooley
Former Member of Congress, Oregon

Lynn Woolsey
Former Member of Congress, California

INDEX